Lecture Notes in Mathematics

Edited by A. Dold and B. Eckmann

648

Nonlinear Partial Differential Equations and Applications

Proceedings of a Special Seminar
Held at Indiana University, 1976–1977

Edited by J. M. Chadam

Springer-Verlag
Berlin Heidelberg New York 1978

Editor
J. M. Chadam
Department of Mathematics
Indiana University
Swain Hall East
Bloomington, IN 47401/USA

AMS Subject Classifications (1970):

ISBN 3-540-08759-1 Springer-Verlag Berlin Heidelberg New York
ISBN 0-387-08759-1 Springer-Verlag New York Heidelberg Berlin

Printed in Germany

Printing and binding: Beltz Offsetdruck, Hemsbach/Bergstr.
2141/3140-543210

Dedicated to Professor Eberhard Hopf

PREFACE

During the 1976-77 academic year a special seminar was held t Indiana University - Bloomington, on Nonlinear Partial Differential quations and Applications. Prominent mathematicians were invited o give pedagogical talks on their current research interests. xcept for Professor Bardos, who presented a three-week mini-course n the Euler equation, most speakers gave one or two 1½ hour ectures in which a topic was developed from its origins to the oint where the open problems could be listed and discussed. This olume consists of the contributions of most of our visiting ecturers. Even a cursory glance at the contents will indicate hy it is natural to dedicate this volume to Professor Eberhard opf. It is a pleasure to have him as a colleague and to have this pportunity to recognize his enormous contribution to both the echniques of Nonlinear Partial Differential Equations and their pplications.

Thanks are due to many of the local people for their ontributions to the program, eppecially Giles Auchmuty, John rothers and Bob Glassey, who gave preparatory lectures, and Elena raboschi, who helped in typing and assembling this volume. Finally, e gratefully acknowledge financial support from the College of Arts nd Sciences and the Office of Research and Graduate Development f Indiana University.

TABLE OF CONTENTS

EULER EQUATION AND BURGER EQUATION - RELATION WITH TURBULENCE.

Claude Bardos
Centre Scientifique et Polytechnique
Université Paris-Nord
Villetaneuse

I. INTRODUCTION

In this paper we are mainly concerned with the existence, uniqueness and regularity during a finite time of solutions of the Euler equation for an ideal two or three dimensional incompressible fluid

(1) $$\frac{\partial u}{\partial t} + u \cdot \nabla u = -\nabla p \quad , \quad \nabla \cdot u = 0$$

plus initial conditions u_0 and boundary conditions on a domain Ω of $\mathbb{R}^n$.

For $n = 3$ the basic result is the following:

Assume that the initial data u_0 is smooth enough, for instance, $u_0 \in C^{1,\alpha}(\Omega)$ (where $C^{1,\alpha}$ denotes the usual Holder space equipped with the natural norm $\|\cdot\|_{1,\alpha}$). Then there exists a unique solution of (1) on the interval $[0,T^*]$ in the space $C^0([0,T^*];C^{1,\alpha}(\Omega))$. T^* is given by the relation $T^* = 1/C\|u_0\|_{1,\alpha}$, where C denotes a constant depending only on Ω.

First, such a result goes back to Lichtenstein [31] and it has been improved by many authors: Ebin and Marsden [13], Kato [22], Temam [38], Foias, Frisch and Temam [14], Bardos and Frisch [7], Benachour [10], Bardos and Benachour [9]. For instance, in [22] it is proved that, when $\Omega = \mathbb{R}^3$, the solution of (1) on $[0,T^*[$, when ε goes to zero, is the limit of the solution u_ε of the corresponding Navier Stokes equation

$$(2) \qquad \frac{\partial u_\varepsilon}{\partial t} - \varepsilon\Delta u_\varepsilon + u_\varepsilon \nabla u_\varepsilon = -\nabla p , \qquad \nabla \cdot u_\varepsilon = 0 .$$

In [13] and [14] it is proved that, if $u_0 \in C^k(\Omega)$ $(k \geq 2)$, then the solution will remain in $C^k(\Omega)$ on the interval $[0,T^*[$. In [9] it is proved that, if u_0 is analytic, then $u(t,\cdot)$ will be analytic as long as it remains in the space $C^{1,\alpha}(\Omega)$.

However, the basic remaining problems are as follows: first, prove that there will be a time T^{**} after which some singularities will appear, and describe these singularities; second, prove that there will be a weak solution defined for all positive times. It is known (Leray [28]) that, for all $\varepsilon > 0$, the problem (2) has a weak solution $u_\varepsilon(t,\cdot)$ defined for all positive times, satisfying the energy estimate

$$(3) \qquad |u_\varepsilon(t,\cdot)|^2 + 2\varepsilon \int_0^b |\nabla u_\varepsilon(s,\cdot)|^2 \, ds \leq |U_0(\cdot)|^2 .$$

(In (3), $|\cdot|$ denotes the $L^2(\Omega)$ norm.) Therefore, using a weak compactness argument, we can define a subfamily still denoted by

u_ε and a function $u \in L^\infty(0,+\infty;L^2(\Omega))$ such that u_ε converges to u in $L^\infty(0,+\infty;L^2(\Omega))$ weak * . Now the main difficulty in passing to the limit in (2) lies in the nonlinear term $u_\varepsilon \cdot \nabla u_\varepsilon$. If $\varphi(\cdot,\cdot)$ is a smooth divergence free vector field defined in $\mathbb{R}_+ \times \Omega$ with compact support, then

(4) $$\int_0^{+\infty}\int_\Omega u_\varepsilon \nabla u_\varepsilon \cdot \varphi \, dxdt = -\int_0^\infty\int_\Omega \sum_{i,j} u_\varepsilon^i u_\varepsilon^j D_j \varphi^i dx\, dt \quad ,$$

and under the weak convergence assumption, one may have

(5) $$\lim_{D'(\Omega\times\mathbb{R}^+)} u_\varepsilon^i u_\varepsilon^j \neq u^i u^j \ .$$

To ensure the equality of the two sides of (5), one would need some pointwise convergence (at least almost everywhere) for $u_\varepsilon(t,x)$. In general, this can be obtained if one has some information on the decay (when $|k| \to \infty$) of the spatial Fourier transform $\hat{u}_\varepsilon(t,k)$ of $u_\varepsilon(t,x)$. Information of this type is linked to the conjecture of Kolmogorov predicting that $\hat{u}_\varepsilon(t,k)$ decays (uniformly in ε) faster than some inverse power of the wave vector. Finally, some information on this decay will in turn give, via a theorem of Frostman [16] (see also Kahane [20]), some information on the Hausdorff measure of the singular support of the solution of (1) after T^{**} .

For $n = 2$ the situation is fairly better, mainly because the curl of the solution is conserved during the motion, and therefore one can prove (Wolibner [40], Schaeffer [37], Kato [21]) that, given smooth initial data, there is a smooth solution defined for all times.

In particular, the solution is analytic when the initial data are analytic (Bardos, Benachour and Zerner [6] or Bardos and Benachour [9]). However, even for $n = 2$, some important problems remain unsolved; for instance, it is not proved that, when ε goes to zero, the solution of the Navier Stokes equation (2) with the natural boundary condition $u_\varepsilon|_{\partial\Omega} = 0$ converges to the solution of the Euler equation. In this case the energy estimate (3) is still valid but, at least up to the present time, one cannot prove that the two sides of (5) are equal. In particular it is known that, due to the boundary layer, u_ε is not uniformly bounded in $H^k(\Omega)$ $(\frac{1}{2} < k)$ (where $H^k(\Omega)$ denotes the usual Sobolev space). Furthermore, the singularities that appear near the boundary of Ω may generally be carried away from this boundary by the fluid. Here again the proof of the convergence is related to a Kolmogorov-type argument and to the study of the set where the curl of u_ε is "big" when ε goes to zero.

Finally, many models have been advanced for the study of the Euler equation, and the simplest one is the Burger-Hopf equation, where all the conjectures can be proved. This paper is organized as follows. In section II the Burger-Hopf equation is proved together with some generalizations of this equation due mainly to Kruckov [24]. The results concerning the Euler equation in three dimensions are proved in section III. Finally, in section IV we give some further information concerning the dimension 2 and in particular we describe Wolibner's method and the way this method is used to prove the persistence of analycity. In section V we conclude with a "rough" description of Kolmogorov's argument.

II. THE BURGER-HOPF EQUATION AND SOME GENERALIZATIONS

In this section $u(\cdot,\cdot)$ will be a scalar valued function defined on $\mathbb{R}^n \times \mathbb{R}_+$, the space variable will be denoted $x = (x_1, x_2, \ldots, x_n)$ and the time variable will be denoted t, as usual. $A(\xi) = (a_1(\xi), a_2(\xi), \ldots, a_n(\xi))$ will denote a continuously differentiable function defined on $\mathbb{R}$ with value in $\mathbb{R}^n$, and for any real valued function u we shall write:

$$\nabla \cdot A(u) = \sum_{i=1}^{n} \frac{\partial}{\partial x_i} (a_i(u)) . \tag{6}$$

We shall study the Cauchy problem for the equation

$$\frac{\partial u}{\partial t} + \nabla \cdot A(u) = 0 , \quad u(\cdot,0) = u_0(\cdot) . \tag{7}$$

When $n = 1$ and $A(u) = u^2/2$, (7) takes the form

$$\frac{\partial u}{\partial t} + u \frac{\partial u}{\partial x} = 0 , \quad u(\cdot,0) = u_0(\cdot) , \tag{8}$$

which is the classical Burger equation.

Assume that u_0 is smooth, and introduce the characteristic curves of the equation (8), i.e., the solution $s \to x(s)$ of the differential equation

$$\frac{dx}{ds} = u(x(s),s) , \quad x(t) = x_t . \tag{9}$$

From equation (8) we deduce that

$$(10) \qquad \frac{d}{ds} u(x(s),s) = 0 \quad ,$$

and therefore u is constant on the lines $x(t) = tu_0(x) + x$. Therefore, the value of the solution at the point (X,T) will be given by solving the equation

$$(11) \qquad X = T u_0(x) + x \quad ,$$

and we will have $u(X,T) = u_0(x)$. Due to the inverse mapping theorem, (11) will have a unique solution x whenever $|\frac{d}{dx}(T u_0(x))| < 1$, e.g.

$$(12) \qquad T < 1/\sup|u_0'(\xi)| \quad .$$

On the other hand, for $T > 1/\sup_{\xi}(-u_0'(\xi))$ there will be two different values x_1 and x_2 such that

$$(13) \qquad T u_0(x_1) + x_1 = T u_0(x_2) + x_2 \; .$$

The two corresponding characteristic lines will merge and therefore there will be no smooth solution for $T > 1/\sup(-u_0'(\xi))$.

We will say that a bounded measurable function is a weak solution of (7) if for any function $\varphi \in \mathcal{D}(\mathbb{R}^n \times \mathbb{R}^+)$ one has

$$(14) \qquad \int_0^\infty \int_{\mathbb{R}^n} (u \cdot \frac{\partial\varphi}{\partial t} + A(u) \cdot \nabla\varphi)\, dx\, dt = 0 \quad .$$

Every smooth solution is a weak solution. Furthermore, if u is piecewise smooth with first order discontinuities along an oriented surface Γ of $\mathbb{R}_+ \times \mathbb{R}$, then

$$(15) \qquad \nu_t(u^- - u^+) + \nu_x \cdot (A(u^-) - A(u^+)) = 0 \quad .$$

(In (15) (ν_t, ν_x) denotes the outward normal to Γ and u^-, u^+ are the values of u inside and outside Γ, respectively.) (15) is generally called the Rankin-Hugoniot condition. One should notice that every weak solution of (8) satisfies the Rankin-Hugoniot condition. On the other hand, the weak solution of (8) is generally not unique (cf. for instance Lax [26, 27]), and in order to insure uniqueness one needs an extra condition which, in the case of the Burger equation, reads

$$(16) \qquad u^- > u^+$$

along any shock curve $\Gamma = (t,x(t))$ of $\mathbb{R}_+ \times \mathbb{R}$.

Relation (16) is called the entropy condition. It has been proved by Hopf [18], Oleinik [35] and others that there is only one weak solution of (16) which satisfies the entropy condition. However, the entropy condition must be satisfied on the shock surfaces which are unknown. Therefore this leads to a "free boundary" problem. As for many "free boundary" problems, one can use a variational formulation and this is Kruckov's main idea, which we describe below.

THEOREM 1. *Assume that* $u_0(\cdot) \in W^{1,1}(\mathbb{R}^n) \cap L^\infty(\mathbb{R}^n)$. *Then there exists*

$$u(t,\cdot) \in C(\mathbb{R}_+;BV(\mathbb{R}^n) \cap L^\infty(\mathbb{R}^n))$$

$$u(t,\cdot) \in C(\mathbb{R}_+;BV(\mathbb{R}^n) \cap L^\infty(\mathbb{R}^n)) \quad *$$

which is a weak solution of the Cauchy problem

$$(17) \qquad \frac{\partial u}{\partial t} + \nabla \cdot A(u) = 0 \quad , \qquad u(\cdot,0) = u_0(\cdot) \quad ,$$

and which satisfies the entropy condition in the following weak sense:

$$(18) \qquad \int_0^\infty \int_{\mathbb{R}^n} |u-k| \frac{\partial \varphi}{\partial t} + \operatorname{sign}(u-k)(A(u)-A(k)) \cdot \nabla\varphi \, dx\, dt \geq 0$$

$$\forall k \in \mathbb{R} \quad \text{and} \quad \varphi \in \mathcal{D}^+(\mathbb{R}_+ \times \mathbb{R}^n)$$

Furthermore the nonlinear operator $S(t)$ *defined by the relation* $u(t,\cdot) = S(t)\,u_0(\cdot)$ *satisfies the relation*

$$(19) \qquad |S(t)u_0^1 - S(t)u_0^2|_{L^1(\mathbb{R}^n)} \leq |u_0^1 - u_0^2|_{L^1(\mathbb{R}^n)}$$

for any pair of initial data $u_0^1, u_0^2 \in W^{1,1}(\mathbb{R}^n) \cap L^\infty(\mathbb{R}^n)$.

Finally, the solution of the Cauchy problem (17) *is the limit, when* ε *goes to zero, of the solution of the perturbed Cauchy problem:*

$$(19) \qquad \frac{\partial u_\varepsilon}{\partial t} - \varepsilon\Delta u_\varepsilon + \nabla \cdot A(u_\varepsilon) = 0 \quad , \qquad u_\varepsilon(\cdot,0) = u_0(\cdot) \quad .$$

Before describing the proof of this theorem we shall make some remarks.

---

* BV denotes the space of function with bounded variation (see Kruckov [24] and Volpert [39]).

REMARK 1. It is easy to see that any function $u \in L^\infty(\mathbb{R}_+ \times \mathbb{R}^n)$ which is a solution of (18) is indeed a weak solution of (17). However, we showed (17) and (18) separately in order to emphasize the relation between (18) and the entropy condition. Indeed, one can prove (see [23] for details) that for any piecewise smooth function u with first order discontinuities on a surface $\Gamma \subset \mathbb{R}^n \times \mathbb{R}_+$, the inequality (18) is equivalent to the following relations:

$$\frac{\partial u}{\partial t} + \nabla \cdot A(u) = 0 \qquad \text{in } \mathcal{D}'(\mathbb{R}^n \times \mathbb{R}_+)$$

and

$$(20) \qquad \nu_t(|u^- - k| - |u^+ - k|) + \nu_x[A(u^-) - A(k)]\,\text{sign}\,(u^- - k)$$
$$- [A(u^+) - A(k)]\,\text{sign}\,(u^+ - k) \geq 0$$

for any $k \in \mathbb{R}$, at any point of Γ.

REMARK 2. Throughout this section we assume for the sake of simplicity that A depends only on u and not on x and t; the general case can be found in [23]. Furthermore, when A does not depend on t, the operator $S(t)$ can be viewed as a nonlinear semigroup. (19) means that $S(t)$ is a contraction semigroup in $L^1(\mathbb{R}^n)$. The link between this type of equation and the theory of nonlinear semigroups in a non-reflexive Banach space is done, for instance, by Crandall [12].

REMARK 3. $S(t)$ is not a nonlinear contraction semigroup in any $L^p(\mathbb{R}^n)$, $p>1$. Otherwise, given $\tau_h^i\varphi(\cdot) = \varphi(x_1,x_2,\ldots,x_i+h,\ldots,x_n)$ one would have

$$(21)\qquad |(\tau_h^i(S(t)u_0) - S(t)u_0)/h|_{L^p} = |(S(t)(\tau_h^i u_0) - S(t)u_0)/h|_{L^p}$$

$$\leq |(\tau_h^i u_0 - u_0)/h|_{L^p} \ .$$

Now, if an estimate of the type (21) is true for $p>1$, we make use of the reflexivity of L^p and thus obtain the relation $S(t)u_0 \in W^{1,p}(\mathbb{R}^n)$ whenever $u_0 \in W^{1,p}(\mathbb{R}^n)$, which is a contradiction because it prevents the appearance of shocks.

REMARK 4. The existence of the solution of the Burger equation, as a limit of the solution of the Burger-Hopf equation, has been obtained directly by Hopf (Hopf [18]) via the following explicit formulas:

$$(22)\qquad u_\varepsilon(x,t) = \frac{\int_{-\infty}^{\infty}\left(\frac{x-y}{t}\right)\exp\left\{-{}^1\!/_{2\varepsilon}\cdot F(x,y,t)\right\}dy}{\int_{-\infty}^{+\infty}\exp\left\{-{}^1\!/_{2\varepsilon}\cdot F(x,y,t)\right\}dy}$$

$$(23)\qquad F(x,y,t) = \frac{(x-y)^2}{2t} + \int_0^y u_0(\eta)\,d\eta \ .$$

Now we shall describe the proof of Theorem 1. We start with the uniqueness and the relation (19).

PROOF. Let u^1 and u^2 be two functions belonging to the space $L^\infty(\mathbb{R}_+ \times \mathbb{R}^n) \cap C(\mathbb{R}_+; BV)$. We take

$$(24) \qquad N = \operatorname{Sup}\Big(\sum_{i=1}^{n} \Big|\frac{da_i}{d\xi}(\xi)\Big|^2\Big)^{\frac{1}{2}}, \qquad |\xi| \le \|u^1\|_\infty + \|u^2\|_\infty .$$

We write the inequalities

$$(25) \qquad \int_0^\infty \int_{\mathbb{R}^n} (|u^1(x,t) - k| \frac{\partial\varphi}{\partial t}(x,y,t,\tau)$$
$$+ \operatorname{sign}(u^1(x,t) - k)(A(u^1) - A(k)) \nabla_x \varphi(x,y,t,\tau) \, dx d\tau \ge 0$$

and

$$(26) \qquad \int_0^\infty \int_{\mathbb{R}^n} (|u^2(y,\tau) - \ell|) \frac{\partial\varphi}{\partial\tau}(x,y,t,\tau)$$
$$+ \operatorname{sign}(u^2(y,\tau) - \ell)(A(u^2) - A(\ell)) \nabla_y \varphi(x,y,t,\tau) \, dy \, d\tau \ge 0$$

with k and ℓ constants, and $\varphi \in \mathcal{D}^+((\mathbb{R}_+ \times \mathbb{R}^n) \times (\mathbb{R}_+ \times \mathbb{R}^n))$. In (25) x,t are the integration variables and y,τ are parameters; in (26) y,τ are the integration variables and x,t are parameters. We replace k with $u^2(y,\tau)$ in (25), and ℓ with $u^1(x,t)$ in (26). We integrate (25) with respect to the variables y,τ and (26) with respect to the variables x,τ . We add the two inequalities and thus obtain the following inequality:

$$(27)\qquad \int_0^\infty\int_0^\infty\int_{\mathbb{R}^n}\int_{\mathbb{R}^n} (|u^1(x,t) - u^2(y,\tau)|)\left(\frac{\partial\varphi}{\partial t} + \frac{\partial\varphi}{\partial\tau}\right) +$$

$$\operatorname{sign}(u^1(x,t) - u^2(y,\tau))(A(u^1) - A(u^2))(\nabla_x\varphi + \nabla_y\varphi)\,dx\,dy\,dt\,d\tau \geq 0 .$$

Now let $f \in \mathcal{D}^+(\mathbb{R}_+ \times \mathbb{R}^n)$ in (27). We shall assume

$$(28)\qquad \varphi(x,y,t,\tau) = f\left(\frac{t+\tau}{2}, \frac{x+y}{2}\right)\delta_h\left(\frac{t-\tau}{2}\right)\delta_h^n\left(\frac{2-y}{2}\right)$$

where δ_h and δ_h^n are smooth nonnegative functions which converge to the Dirac distribution in $\mathbb{R}$ and $\mathbb{R}^n$ when h goes to zero. Letting h go to zero we finally obtain (see [24] for details):

$$(29)\qquad \int_0^\infty\int_{\mathbb{R}^n} (|u^1(x,t) - u^2(x,t)|)\frac{\partial f}{\partial t} +$$

$$\operatorname{sign}(u^1 - u^2)(A(u^1) - A(u^2))\cdot\nabla f)\,dx\,dt \geq 0 .$$

Now, let

$$(30)\qquad f_h^R(x,t) = (\alpha_h(t+2h) - \alpha_h(t-T))(1-\alpha_h(|x| + Nt - R)) .$$

In (30) N is defined by (24), R is a positive number that will eventually go to infinity and $\alpha_h(s) = \int_{-\infty}^s \delta_h(\xi)\,d\xi$. We have

$$(31)\qquad \operatorname{supp} f \subset \left\{(x,t)\,|\,h \leq t \leq T+h, |x| \leq Nt + R + h\right\}$$

Now in (29) we replace f with f_h^R and we let R go to infinity. Making use of the inequality

$$(32)\qquad |\mathrm{sign}\,(u^1 - u^2)(A(u^1) - A(u^2)) \cdot \frac{x}{|x|}| \le |A(u^1) - A(u^2)|$$

$$\le N|u^1 - u^2|$$

(Mean value theorem!)

we obtain

$$(33)\qquad \int_0^\infty \int_{\mathbb{R}^n} |u^1(x,t) - u^2(x,t)|(\alpha_h^1(t+2h) - \alpha_h^1(t-T))\,dx\,dt \ge 0 \quad .$$

Finally, letting h go to zero, we deduce from the inequality (33) the inequality (19) and the uniqueness of the solution.

REMARK 5. The same method is used in [24] to prove that the solution propagates with finite speed.

Now we shall show that the solution $u_\varepsilon(x,t)$ of the perturbed problem (19) converges to a function u which is a solution of (18). We multiply the equation

$$(19)\qquad \frac{\partial u_\varepsilon}{\partial t} - \varepsilon\Delta u_\varepsilon + \nabla \cdot A(u_\varepsilon) = 0$$

by u_ε, and integration by part yields the classical estimate:

$$(34)\qquad \tfrac{1}{2}|u_\varepsilon(\cdot,t)|^2 + \varepsilon\int_0^t |\nabla u_\varepsilon(\cdot,s)|^2 ds = \tfrac{1}{2}|u_0(\cdot)|^2 \quad .$$

Furthermore, using the maximum principle we obtain

$$(35) \qquad |u_\varepsilon|_{L^\infty(\mathbb{R}_+ \times \mathbb{R}^n)} \le |u_0(\cdot)|_{L^\infty(\mathbb{R}^n)} \quad .$$

Finally we notice that we have

$$(36) \qquad \int_{\mathbb{R}^n} -\Delta\left(\frac{\partial u}{\partial x_i}\right) \operatorname{sign}\left(\frac{\partial u}{\partial x_i}\right) dx \ge 0$$

and

$$(37) \qquad \int_{\mathbb{R}^n} \frac{\partial}{\partial x_i} (\nabla \cdot A(u)) \operatorname{sign}\left(\frac{\partial u}{\partial x_i}\right) = 0$$

for any smooth function u . From (36) and (37) we deduce the crucial estimate:

$$(38) \qquad \frac{d}{dt}\left(\int_{\mathbb{R}^n} \left|\frac{\partial u_\varepsilon}{\partial x_i}\right| dx\right) \le 0 \quad .$$

The estimate (38) shows that u_ε is bounded in $L^\infty(\mathbb{R}_+;W^{1,1}(\mathbb{R}^n))$, and therefore that a subsequence still denoted u_ε will converge, when ε goes to zero, to a function belonging to $L^\infty(\mathbb{R}_+;BV)$. From (38) we also deduce that a subsequence u_ε converges almost everywhere (use the classical compactness theorems, e.g., Lions [30], p. 11). Therefore, using the Lebesgue theorem and the continuity of A , we can prove that $\nabla \cdot A(u_\varepsilon)$ goes to $\nabla \cdot A(u)$ in $\mathcal{D}'(\mathbb{R}_+ \times \mathbb{R}^n)$. This shows that u , the weak limit of u_ε , is a weak solution of (17). Finally, we shall obtain the relation (18). We introduce a smooth convex function, Φ , and multiply the equation (19) by $\Phi'(u_\varepsilon)\varphi$ $(\varphi \in \mathcal{D}^+(\mathbb{R}^+ \times \mathbb{R}^n)$.

An integration by part yields

$$(39)\qquad \iint\Big(-\Phi(u_\varepsilon)\frac{\partial\varphi}{\partial t} + \varepsilon|\nabla u_\varepsilon|^2\,\Phi''(u_\varepsilon)\varphi$$

$$+\ \varepsilon\,\nabla u_\varepsilon\cdot\nabla\varphi\,\Phi'(u_\varepsilon)$$

$$+\ \varphi\cdot\sum_{i=1}^{n}\frac{\partial}{\partial x_i}\Big(\int_k^{u_\varepsilon} a_i'(s)\Phi'(s)\,ds\Big)\Big)\,dx\,dt = 0\ .$$

(k is any constant and $\int_k^{\xi} a_i'(s)\Phi'(s)\,ds$ is a primitive of $a_i'(\xi)\Phi'(\xi)$.) Since φ is positive and Φ is convex, we have

$$\varepsilon|\nabla u_\varepsilon|^2\,\Phi''(u_\varepsilon)\varphi \geq 0\ .$$

Due to estimate (34) we have:

$$\lim_{\varepsilon\to 0}\iint \varepsilon\,\nabla u_\varepsilon\cdot\nabla\varphi\,\Phi'(u_\varepsilon)\,dx\,dt = 0\ .$$

Therefore, letting ε go to zero and using the almost everywhere convergence of u_ε , we deduce from (39) that

$$(40)\qquad \int_0^\infty\int_{\mathbb{R}^n} \Phi(u)\,\frac{\partial\varphi}{\partial t} + \sum_{i=1}^{n}\Big(\int_k^{u} a_i'(s)\Phi'(s)\,ds\Big)\frac{\partial\varphi}{\partial x_i}\,dx\,dt \geq 0$$

for any convex function Φ and any smooth function $\varphi\in\mathcal{D}^+(\mathbb{R}_+\times\mathbb{R}^n)$. Finally, we introduce a sequence Φ_h of convex functions such that

$$(41)\qquad \lim_{h\to 0}\Phi_h(\xi) = |\xi-k|,\ \lim_{h\to 0}\Phi_h'(\xi) = \operatorname{sign}(\xi-k)$$

and we deduce (18) from (40) by replacing ϕ with ϕ_h and letting h go to zero.

III. REGULAR SOLUTIONS OF THE EULER EQUATION IN THREE DIMENSIONS

We shall consider the Euler equation for an inviscid incompressible fluid in an open set Ω of $\mathbb{R}^3$. In the introduction we wrote the equation

$$(41) \qquad \frac{\partial u}{\partial t} + u \cdot \nabla u = -\nabla p \;, \qquad \nabla \cdot u = 0 \;.$$

For the sake of simplicity, we will only consider the two following cases:

(a) Ω is bounded, simply connected with a smooth boundary $\partial\Omega$. $\nu = (\nu_1, \nu_2, \nu_3)$ will denote the outward normal to $\partial\Omega$. In this case the boundary condition is

$$(42) \qquad \nu \cdot u = 0 \quad \text{on} \quad \partial\Omega \;.$$

(b) $\Omega = \mathbb{R}^3$ (in this case there is no boundary condition!).

Following Bardos and Frisch [7] we shall show that the regularity of the solution is mainly governed by the Holder norm of the vorticity. Although there is no proof of the appearance of singularity, there are several reasons to believe that singularities will appear at a time T^*, which is of the order of the inverse of the maximum of the modulus of the initial vorticity. First, we have seen in section

II that this is the case for the Burger equation. Secondly, there is a "vortex-stretching" heuristic argument: we take the curl of (1) in order to obtain (with $\omega = \nabla \wedge u$ and $\frac{D}{Dt}$ denoting the Lagrangian derivative)

$$(43) \qquad \frac{D\omega}{Dt} = \frac{\partial\omega}{\partial t} + u \cdot \nabla\omega = \omega \cdot \nabla u \quad .$$

Notice that the vorticity is just the antisymmetric part of the velocity gradient ∇u ; if we tentatively identify ω and ∇u and discard the vector and tensor indices we obtain $\frac{D\omega}{Dt} = \omega^2$, which implies that ω should blow up at the time $T^* = 1/|\sup \nabla \wedge u_0(\cdot)|$.

We shall show that there is a solution on an interval of this type; that, in case (b), this time does not depend on the total energy of the fluid; and we shall finally show that there is no loss of regularity as long as the curl remains bounded. More precisely, we shall show that, if $u_0 \in C^\infty(\Omega)$, the solution will belong to $C^\infty(\Omega)$ (and that, if u_0 is analytic, the solution will be analytic) as long as $\nabla \wedge u$ remains bounded.

We note that $C^{k,\alpha}(\Omega)$ denotes the usual* Hölder spaces, that $\nabla\cdot$, ∇ , $\nabla\wedge$, and $u \cdot \nabla$ denote the usual operators, divergence, gradient curl and $\sum_{i=1}^{3} u_i \frac{\partial u}{\partial x_i}$; in many cases the symbol $\sum$ will be omitted and $\frac{\partial}{\partial x_i}$ replaced by D_i . By $C_\sigma^{k,}(\Omega)$ we shall denote the space

$$C_\sigma^{k,\alpha}(\Omega) = \left\{ u \in C^{k,\alpha}(\Omega),\ \nabla \cdot u = 0\ ,\ \nu(\cdot) \cdot u(\cdot)|_{\partial\Omega} = 0 \right\} \ .$$

---

* We shall denote its natural norm by $\|\cdot\|^{k,\alpha}$.

Finally, to study the analyticity of the solution we shall introduce the space $\mathbb{C}^3 = \{(x_1 + iy_1, x_2 + iy_2, x_3 + iy_3)\}$. In case (a), for $x \in \mathbb{R}^3$, we denote the distance of x to $\partial\Omega$ by $d(x)$, and we introduce a strictly positive number β with the following properties. Restricted to the set $d^{-1}([0,\alpha[)$, the function d is smooth and there is for every $x \in d^{-1}([0,\alpha[)$ a unique point, $p(x) \in \partial\Omega$ such that $d(x) = |x - p(x)|$. Finally, the mapping $x \to p(x)$ is smooth. We shall say that a subset $\mathcal{O}(\lambda,\mu)$ (depending on two parameters λ and μ) of $\mathbb{C}^3$ is an admissible complex neighborhood of Ω if we have

$$(44) \qquad \mathcal{O}(\lambda,\mu) = \Big\{ (x + iy) \in \mathbb{C}^3 \,|\, x \in \Omega, \ |y| \le \mu$$

$$|y| < \lambda \ d(\lambda) \quad \text{whenever} \quad x \in d^{-1}([0,\alpha[) \Big\} \ .$$

We shall assume that we have $0 < \lambda < 1$ and $\mu / a < \lambda$. Also, we shall write $\mathcal{O}(\lambda(t),\mu(t)) = \mathcal{O}(t)$ whenever λ and μ depend on a parameter t . For any admissible complex neighborhood of Ω , $\mathcal{O}(\lambda,\mu)$, $B(k,\alpha,\mathcal{O})$ will denote the functions holomorphic in $\mathcal{O}$ with derivatives up to the order k bounded in $\overline{\mathcal{O}}$ and the Hölder quotient of order $k+\alpha$: $(D^k v(z+h) - D^k v(z))/|h|^\alpha$ bounded in $\overline{\mathcal{O}}$. $\| \cdot \|_{\mathcal{O}}^{k,\alpha}$ will be the natural norm on the Banach space $B(k,\alpha,\mathcal{O})$. We shall make use of the following proposition.

PROPOSITION 1. *Assume that* Ω *is a simply connected bounded set with smooth boundary* $\partial\Omega$. *Then for any divergence free vector field* w *defined on* Ω *belonging to the space* $C^{0,\alpha}(\Omega)$ *there exists a unique vector field* u *which is the solution of the elliptic boundary value problem*:

(45) $\quad \nabla \wedge u = w \,, \; \nabla \cdot u = 0 \,, \quad u \cdot \nu|_{\partial\Omega} = 0 \;.$

Furthermore for any $k \geq 1$, *there exists a constant* C_k *such that*

(46) $\quad \|u\|^{k+1,\alpha} \leq C_k \|w\|^{k,\alpha}$ *whenever* $w \in C^{k,\alpha}(\Omega)$.

Finally, whenever $w \in B(k,\alpha,\mathcal{O})$, *the solution of* (45) *is the restriction to* Ω *of a function* (still denoted u) *belonging to the space* $B(k+1,\alpha,\mathcal{O})$; *hence*

(47) $\quad \|u\|_{\mathcal{O}}^{k+1,\alpha} \leq C_k(\mathcal{O}) \|w\|_{\mathcal{O}}^{k,\alpha}$.

($C_k(\mathcal{O})$ is an increasing function of $\mathcal{O}$ $(\hat{\mathcal{O}} \supset \mathcal{O} \Rightarrow C_k(\hat{\mathcal{O}}) \geq C_k(\mathcal{O}))$.)

The first part of the statement of Proposition 1 is a classical result on ellipticity (see, for instance, Agmon Douglis & Nirenberg [1]); the second, concerning analyticity, is possibly slightly sharper than the usual results (see Morrey [32]) and can be found in Bardos and Benachour [9], or deduced from a paper of Kahane [19] .

In the case where $\Omega = \mathbb{R}^3$ we shall make use of the following results.

PROPOSITION 2. *For any* k *there is a constant* C_k *such that, for any* $u \in C_\sigma^{k+1,\alpha}(\mathbb{R}^3)$,

$$(48) \qquad \|u\|^{k+1,\alpha} \le C_k\left\{\|\nabla \wedge u\|^{k,\alpha} + \|u\|^{k,\alpha}\right\} .$$

This result is easy to deduce from Proposition 1; an explicit integral representation can also be used.

Due to propositions 1 and 2, the spaces $C_\sigma^{k,\alpha}(\Omega)$ in case (a) and $C_\sigma^{k,\alpha}(\mathbb{R}^3)$ in case (b) will be equipped with the following equivalent norms, respectively:

$$(49) \qquad |||\, u\, |||_{k,\alpha} = \|\nabla \wedge u\|_{k-1,\alpha},\quad |||\, u\, |||_{k-1,\alpha} = \left\{\|\nabla \wedge u\|_{k-1,\alpha} + \|u\|_{k-1,\alpha}\right\}.$$

Finally, the space $B_\sigma(k,\alpha,\Theta) = \left\{u \in B(k,\alpha,\Theta),\ \nabla \cdot u = 0,\ u \cdot \nu|_{\partial\Omega} = 0\right\}$ will be equipped with the following equivalent norm:

$$(50) \qquad |||\, u\, |||_{k,\alpha,\Theta} = \|\nabla \wedge u\|_{k-1,\alpha,\Theta} .$$

THEOREM 2. *Let* Ω *be a bounded simply connected open set of* $\mathbb{R}^3$. *Then there exists a constant* C_1 *depending only on* Ω *and* α $(0 < \alpha < 1)$ *such that for every* $u_0 \in C_\sigma^{1,\alpha}$ *there exists a unique function* $u \in C(-T^*, T^*; C_\sigma^{1,\alpha})$ *which is a solution of the Euler equation and which for* $t = 0$ *is equal to* u_0 . *Furthermore,* T^* *is given by the relation* $T^* = \left(C_1 |||\, u_0 |||_{1,\alpha}\right)^{-1}$ *and* u *satisfies the estimate*

$$(51) \qquad |||\, u(t)\, |||_{1,\alpha} \le |||\, u_0\, |||_{1,\alpha} \frac{T^*}{T^* - |t|} .$$

THEOREM 3. *Let* $\mathcal{O}(\lambda, \mu)$ *be an admissible complex neighborhood of* Ω (bounded and simply connected). *Assume that* u_0 *is a real function which is the restriction to* Ω *of a function* (still denoted u_0) *belonging to* $B_\sigma(1,\alpha,\mathcal{O}(\lambda,\mu))$. *Let* $M = |||u_0|||^{\mathcal{O}}_{0,\alpha}$, $M(t) = M/(1-kM|t|)$ *where* k *is a constant which is chosen large enough. Then for* $|t| < (kM)^{-1}$ *the solution of the Euler equation is the restriction to* Ω *of an analytic function* $u(t)$ *which belongs to the space* $B_\sigma(1,\alpha,\mathcal{O}(t))$. $\mathcal{O}(t)$ *is given by the relation*

$$\mathcal{O}(t) = \mathcal{O}(\lambda(1 - kMt), \mu(1 - kMt)) , \tag{52}$$

and we have the a priori estimate

$$|||u(t,\cdot)|||^{\mathcal{O}(t)}_{0,\alpha} \le M/(1 - kMt) . \tag{53}$$

THEOREM 4. *Let* Ω *be a simply connected open set of* $\mathbb{R}^3$ *and* $u(x,t)$ *a real function which is a solution of the Euler equation in* $\Omega \times [0,T[$. *Assume that* u *is continuously differentiable in* $\Omega \times [0,T[$ *and that* $u(0,\cdot)$ *is the restriction to* Ω *of an analytic function which belongs to* $B_\sigma(1,\alpha,\mathcal{O})$ ($\mathcal{O}$ is an admissible complex neighborhood of Ω). *Then for any* $t (0 < t < T)$, $u(t,\cdot)$ *is analytic in the variable* x .

REMARK 6. As long as the solution is smooth one can change t into $-t$; therefore, the proof will be given only for $t > 0$. Theorem 3 is a classical Cauchy-Kowalevski theorem: in the statement as well the proof it uses a scale of spaces $\left(B(1,\alpha,\mathcal{O}(t))\right)$ as was introduced by Ovsjannikov (Ovsjannikov [36]; see also Nirenberg [33], Baouendi & Goulaovic [2] and Nishida [34]). Theorem 4 uses the hyperbolic character of the problem and generally it is not possible to prove that the analytic regularity persists as long as does the C^∞ regularity.

Since the solutions are smooth the uniqueness is always easy to prove. The existence of the solution will be done by approximation: we shall construct some sequences $u^n(t,\cdot)$ and prove that they are uniformly bounded in n, for $0<t<T$. The rest of the proof is then classical.

PROOF OF THEOREM 2. We introduce the sequence u^n, w^n which is defined via the following formulas:

$$(54)\qquad u^0(x,t) = u_0(x)\ ,\ w^0(x,t) = \nabla\wedge\cdot u_0(x)\ .$$

$$(55)\qquad \frac{\partial w^{n+1}}{\partial t} + u^n\cdot\nabla w^{n+1} = w^{n+1}\cdot\nabla u^n\ ,\ w^{n+1}(0,\cdot) = \nabla\wedge u_0$$

$$(56)\qquad \nabla\wedge u^{n+1} = w^{n+1}\ ,\ \nabla\cdot u^{n+1} = 0\ ,\ u^{n+1}\cdot\nu|_{\partial\Omega} = 0\ .$$

Assume that we have $\nabla\cdot w^n = 0$, and let

$$(57)\qquad \|w^n(\cdot,t)\|_{0,\alpha} \le |||u_0|||_{1,\alpha}\frac{T^*}{T^*-t}\qquad (0<t<T^*)\ .$$

We can obtain u^n via the formula (56) (at the order n) . Next we introduce the solutions of the ordinary differential equation

$$(58)\qquad x'(s) = u^n(x(s),s) \ , \quad x(t) = x \in \Omega \ .$$

It is easy to see (cf. Bourguignon & Brezis [11]) that the solution of (58) is well defined and belongs to Ω for any $0 < t$. The curve $s \to x(s)$ defines what is usually called the Lagrangian coordinates. The equation (55) can then be written in the following form:

$$(59)\qquad \frac{d}{dt} w^{n+1}(x(t),t) = (w^{n+1} \cdot \nabla u^n)(x(t),t) \ .$$

From (59) one deduces (see Bardos & Frisch [7] for details)

$$(60)\qquad \|w^{n+1}(x(t),t)\|_{0,\alpha} \le \|w_0\|_{0,\alpha} + \int_0^t (\|w^{n+1} \cdot \nabla u^n\|_{0,\alpha} + \|\nabla u^n\|_{L^\infty(\Omega)} \|w^{n+1}\|_{0,\alpha}) \, ds \ .$$

Finally one uses (46) to obtain the estimate

$$(61)\qquad \|w^{n+1}(x(t),t)\|_{0,\alpha} \le \|w_0\|_{0,\alpha} + C_1 \int_0^t \|w^n\|_{0,\alpha} \|w^{n+1}\|_{0,\alpha} \, ds \ ,$$

and by Gronwall's lemma one deduces from the relation (57) that

$$\|w^{n+1}(\cdot\, , t)\|_{0,\alpha} \le |||\, u_0 \,|||_{1,\alpha} \frac{T^*}{T^* - t}$$

holds true.

To prove that, if $u_0 \in C^{k,\alpha}_\sigma(\Omega)$, the solution remains in the same space for $t < T^*$, one may differentiate up to the order ℓ $(|\ell| \le k)$ in equation (55), thus obtaining

$$(62) \qquad \frac{\partial}{\partial t} D^\ell w^{n+1} + u^n \nabla D^\ell w^{n+1} = D^\ell w^{n+1} \nabla u^n + g \ .$$

By using the Leibnitz formula one can show that all the terms appearing in g are of an order strictly less than $|\ell| + 1$. Therefore with the formula (46) one can prove by iteration that $D^\ell u^n$ is uniformly in the space $C^{1,\alpha}(\Omega)$ for $t < T^*$. $\left(\ell = (\ell_1, \ell_2, \ell_3) \right.$ is a multiinteger; $D^\ell u = \frac{\partial^{\ell_1}}{\partial x_1} \frac{\partial^{\ell_2}}{\partial x_2} \frac{\partial^{\ell_3}}{\partial x_3} > |\ell| = \ell_1 + \ell_2 + \ell_3 \left. . \right)$ Since this estimate is uniform with respect to n, it is easy to complete the proof of Theorem 2.

PROOF OF THEOREM 3. To prove Theorem 3 one follows the same path using the extension to the complex domain. Let

$$\Sigma = \left\{ (z,t) \mid 0 < t < T^*, \quad z \in \mathcal{O}(t) \right\}$$

where $\mathcal{O}(t)$ is defined by the relation (52). In Σ a sequence of holomorphic functions is defined via the following iteration formulas:

$$(63) \qquad u^n(t) \in B_\sigma(1,\alpha,\mathcal{O}) \ , \quad \nabla \wedge u^n = w^n \ .$$

$$(64) \qquad \frac{\partial w^{n+1}}{\partial t} + (u^n \cdot \nabla) w^{n+1} = (w^{n+1} \cdot \nabla) u^n \ , \quad w^n(\cdot,0) = \nabla \wedge u_0 \ .$$

The relations (63) and (64) determine a sequence (u^n, w^n) which satisfies the following relations:

$$(65) \qquad \nabla \cdot w^n = 0 \ , \quad \|w^n(t,0)\|_{0,\alpha}^{\mathcal{O}(t)} \leq M(t) \quad .$$

Assume that the relations (65) are true up to the order n. One can define u^n by means of the formula (63). Next using (47) one can show that

$$(66) \qquad \sup_{z \in \mathcal{O}(t)} |\nabla u^n(t,z)| \leq C(\mathcal{O}(0))\, M/t) \quad .$$

Using the mean value theorem, given that $u^n(t,z)$ is real when $z \in \Omega$ and tangent to $\partial\Omega$ when $z \in \partial\Omega$, one can show that the solution of the differential equation

$$(67) \qquad z'(s) = u^n(z(s),s) \ , \ z(t) = z$$

is defined for any $0 \leq s \leq t$ and belongs to Σ, whenever $z \in \mathcal{O}(t)$ $(0 < t < T^*)$.

Therefore one can solve (66) by writing it in the form

$$(68) \qquad \frac{d}{ds} w^{n+1}(z(s),s) = \left((w^{n+1} \cdot \nabla)u^n\right)(z(s),s)$$

(see Bardos & Benachour [9] for details). Since $z(s)$ remains in $\mathcal{O}(s)$ for any s $(0 \leq s \leq t)$, one can prove the a priori estimate

(69) $$\|w^{n+1}(t,\cdot)\|_{0,\alpha}^{\mathcal{O}(t)} \leq \|w^{n+1}(0,\cdot)\|_{0,\alpha}^{\mathcal{O}}$$

$$+ C\int_0^t \|\nabla u^n\|_{L^\infty(\mathcal{O}(s))} \|w^{n+1}\|_{0,\alpha}^{\mathcal{O}(s)} + \|w^{n+1}\varphi - \nabla u^n\|_{0,\alpha}^{\mathcal{O}(s)} ds \quad .$$

Finally, by using again Gronwall's Lemma, one may deduce from (65) that $\|w^{n+1}(t,0)\|_{0,\alpha}^{\mathcal{O}(t)} \leq M(t)$.

PROOF OF THEOREM 4. Let T^{**} denote the upper bound of the number T' with the following properties. For any t , $0<t<T'$, there exists a complex admissible neighborhood $\mathcal{O}(t)$ of Ω such that $u(\cdot\,,t)$ is the restriction of a function (still denoted u) belonging to the space $B(1,\alpha,\mathcal{O}(t))$. From Theorem 3 we already know that $T^{**} > 0$, and we shall show that $T^{**} = T$. By eventually shrinking the neighborhood $\mathcal{O}(t)$, one may assume that, for any $t \in [0,T[$, $\mathcal{O}(t) \subset \mathcal{O}(0)$, and therefore the same constant will appear in formula (47) for all the neighborhood $\mathcal{O}(t)$. Let $T^{**} < T$ and $T^{**} < T' < T$. On the interval $[0,T']$, the first and second derivatives of u are uniformly bounded in the real domain by a constant $P>0$ (use, for instance, the result by Foias, Frisch & Temam [14]). Let $\varepsilon>0$, $u(T^{**}-\varepsilon)$ is the restriction on Ω of an holomorphic function $u(T^{**}-\varepsilon)$ defined in some admissible complex neighborhood of Ω , $\mathcal{O}(T^{**}-\varepsilon)$. In Ω the first and second derivatives of $u(T^{**}-\varepsilon,\cdot)$ are uniformly bounded by P . Therefore, using the continuity of $u(T^{**}-\varepsilon,\cdot)$, one may assume with a complex neighborhood $\tilde{\mathcal{O}}(T^{**}-\varepsilon) \subset \mathcal{O}(T^{**}-\varepsilon)$ (chosen small enough) the following estimate:

(70) $$\|\nabla \wedge u(T^{**} - \varepsilon , \cdot)\|_{0,\alpha}^{\tilde{\mathfrak{G}}(T^{**}-\varepsilon)} \leq (P+1) \quad .$$

Finally, changing t to $t + T^{**} - \varepsilon$ and using Theorem 3, one can see that $u(t, \cdot)$ is the restriction on the real domain of a function which is analytic in x for $T^{**} - \varepsilon \leq t < (T^{**} - \varepsilon) + 1/k(P+1)$ (where k is the constant which appears in (53)). With $\varepsilon < 1/k(P + 1)$ it is shown that T^{**} is not the upper bound of the T' such that $u(t, \cdot) \in B(1,2,\mathfrak{G}(t))$ for $0 < t < T'$, and the proof is complete.

REMARK 7. The analyticity with respect to t is not described here; the results and methods are analogous and can be found in Benachour [10].

Finally, we shall study the existence and uniqueness of the solution in the whole space. As we have noted already, the important point is that we shall prove the existence of the solution up to a time T^* which is independent of the total energy of the initial data. This will allow us to consider also initial data with infinite total energy, which is the case with homogeneous turbulence.

Let $\varphi(x) \in \mathcal{D}(\mathbb{R}^3)$, and assume that $\varphi(x) = 1$ in a neighborhood of the origin. We state $\tilde{\varphi}(x) = 1 - \varphi(x)$, and we introduce the operator $(u,v) \to F_\varphi(u,v)$ defined by the formulas

(71) $$F_\varphi(u,v) = \nabla \int_{\mathbb{R}^3} D_i \left(\frac{\varphi(x-y)}{4\pi|x-y|}\right)(u_j D_j v_i)(y)\, dy$$

$$+ \int_{\mathbb{R}^3} D^2_{ij} \nabla \left(\frac{\tilde{\varphi}(x-y)}{4\pi|x-y|}\right)(v_i u_j)(y)\, dy \quad .$$

The integral

$$\int_{\mathbb{R}^3} D_i\left(\frac{\varphi(x-y)}{4\pi|x-y|}\right)(u_j D_j v_i)(y)\, dy$$

is a convolution with a function having compact support and a singularity at the origin behaving like $|x|^{-2}$. And the integral

$$\int_{\mathbb{R}^3} D^2_{ij}\left(\nabla\left(\frac{\tilde{\varphi}(x-y)}{4\pi|x-y|}\right)\right)(v_i u_j)(y)\, dy$$

is a convolution with a smooth function which behaves asymptotically like $|x|^{-4}$ and therefore belongs to the space $L^1(\mathbb{R}^3) \cap C^\infty(\mathbb{R}^3)$. (Taking three derivatives of $|x|^{-1}$ introduces a term of the order $|x|^{-4}$.) Therefore, it is easy to see (use the classical results on Schauder estimates, Ladyzenskaia & Uralceva [25], p. 116) that $(u,v) \to F_\varphi(u,v)$ is a bilinear continuous map defined in $C^{0,\alpha}(\mathbb{R}^3) \times C^{1,\alpha}(\mathbb{R}^3)$ with value in $C^{0,\alpha}(\mathbb{R}^3)$. Furthermore, F_φ is independent of the choice of φ and therefore the symbol φ will be omitted. Finally, F is a gradient and whenever $\nabla \cdot v = 0$ one has the relation

$$\nabla \cdot F(u,v) = D_i u_j D_j v_i + R_j(v_j \cdot (\nabla \cdot u)) \tag{72}$$

where R_j is a linear bounded operator defined in $C^{0,\alpha}(\mathbb{R}^3)$ with value in $C^{0,\alpha}(\mathbb{R}^3)$.

THEOREM 5. *There exists a constant* C_2 (depending on α) *such that for any* $u_0 \in C^{1,\alpha}(\mathbb{R}^3)$ $(\nabla \cdot u_0 = 0)$ *there is a unique function* $u \in C(-T^*; T^*; C^{1,\alpha}(\mathbb{R}^3))$, $T^* = (C|||u_0|||)^{-1}$ *with the following properties*:

(i) u *is a solution of the equation*

$$\frac{\partial u}{\partial t} + u \cdot \nabla u = F(u,u) \; , \; \nabla \cdot u = 0 \quad . \tag{73}$$

(ii) *On the interval* $]-T^*, T^*[$, u *satisfies the estimate*

$$|||u(t)|||_{1,\alpha} \le |||u_0|||_{1,\alpha} \frac{T^*}{T^* - |t|} \quad . \tag{74}$$

(iii) *When* $u_0 \in C^{1,\alpha}(\mathbb{R}^3)$ *belongs to* $L^2(\mathbb{R}^3)$ (finite energy) *the solution of* (73)-(74) *coincides with the usual solution of the Euler equation*:

$$\frac{\partial u}{\partial t} + u \cdot \nabla u = -\nabla p \quad (\nabla p \in L^2) \; , \quad u(x,0) = u_0(x) \; .$$

REMARK 7. The norm $||| \cdot |||_{1,\alpha}$ is defined by the right-hand side of (46). Some results of the same type may be obtained when Ω is the exterior of a bounded closed set, the natural boundary condition then being $u \cdot \nu|_{\partial\Omega} = 0$. The C^∞ and analytic regularity may also be studied; they are omitted here (see Bardos & Benachour [9] for details).

REMARK 8. As it will be described in the next section, global regularity holds in dimension two when the open set Ω is bounded or when the total energy of the initial data is finite. It is an open question whether the global regularity (in time) holds for the solution of a problem of the type (73)-(74) in dimension two when the total energy of the initial data is not finite.

REMARK 9. When u is of finite total energy the Euler equation determines p (up to a constant) by the formula

$$(75) \qquad -\nabla p = D_i u_j D_j u_i \quad .$$

In this case ∇p belongs to $L^2(\mathbb{R}^3)$.

However, in the general case the operator F takes its value in $C^{0,\alpha}(\mathbb{R}^3)$ and not in $L^2(\mathbb{R}^3)$. Therefore the relation

$$-\Delta p = \nabla\left(\frac{\partial u}{\partial t} + u \nabla u\right) = D_i u_j D_j u_i$$

defines ∇p up to a constant vector. Hence the need to introduce the operator F to specify the choice of ∇p .

PROOF OF THEOREM 5. We shall use the iteration scheme defined by the relation

$$(76) \qquad \frac{\partial u^{n+1}}{\partial t} + u^n \nabla u^{n+1} = F(u^{n+1}, u^n), u^{n+1}(0, \cdot) = u_0(\cdot) \quad ,$$

and assume that $\nabla \cdot u^n = 0$ and that u^n satisfies the a priori estimate.

$$(77) \qquad |||u^n(t)|||_{1,\alpha} \le |||u_0|||_{1,\alpha} \frac{T^*}{T^* - |t|} .$$

Once again one introduces the Lagrangian coordinates, i.e., the solution of the equation

$$x'(s) = u^n(x(s),s) , \qquad x(t) = x .$$

Equation (76) then becomes

$$\frac{d}{dt} u^{n+1}(x(t),t) = F(u^{n+1},u^n)(x(t),t) .$$

Now, taking the divergence of (76) and using (72), one may obtain for the divergence of u^{n+1} the relation

$$(78) \qquad \frac{\partial}{\partial t}(\nabla \cdot u^{n+1}) + (u^n \cdot \nabla)(\nabla \cdot u^{n+1}) = R_j(u^n \cdot (\nabla \cdot u^{n+1})) .$$

From (78) we deduce that $\nabla \cdot u^{n+1}(t,\cdot) \equiv 0$. Finally, one takes the curl of (76) and by using the fact that F is a gradient the following relation is obtained:

$$(79) \qquad \frac{\partial}{\partial t}(\nabla \wedge u^{n+1}) + u^n{}_\nabla(\nabla \wedge u^{n+1}) = B(u^n , u^{n+1}) .$$

In (79), B is a first order bilinear differential operator which satisfies the relation

$$\|B(u^n, u^{n+1})\|_{0,\alpha} \leq C\|u^n\|_{1,\alpha} \|u^{n+1}\|_{1,\alpha} \quad . \tag{80}$$

The relation (79) can also be written in Lagrangian coordinates, in the form

$$\frac{d}{dt}(\nabla \wedge u^{n+1})(x(t),t) = B(u^n, u^{n+1})(x(t),t) \quad . \tag{81}$$

From (77) and (81) one may deduce the two following estimates:

$$\|u^{n+1}(t)\|_{0,\alpha} \leq \|u^{n+1}(0)\|_{0,\alpha} + C\int_0^t (|\nabla u^n|_{L^\infty(\mathbb{R}^3)} \|u^{n+1}\|_{0,\alpha} + \|F(u^n,u^{n+1})\|_{0,\alpha})\, ds \tag{82}$$

$$\|\nabla \wedge u^{n+1}(t)\|_{0,\alpha} \leq \|\nabla \wedge u_0\|_{0,\alpha} + \int_0^t (|\nabla u^{n+1}|_{L^\infty(\mathbb{R}^3)} \|\nabla \wedge u^{n+1}\|_{0,\alpha} + \|B(u^n,u^{n+1})\|_{0,\alpha})\, ds \tag{83}$$

Adding (82) and (83), and using (48), one obtains the relation

$$|||u^{n+1}(t)|||_{1,\alpha} \leq |||u_0|||_{1,\alpha} + C\int_0^t |||u^{n+1}|||_{1,\alpha} |||u^n|||_{1,\alpha}\, ds \tag{84}$$

Once again, one deduces from Gronwall's Lemma that the estimate (77) is valid at the order $n+1$. The rest of the proof is classical and left to the reader.

IV. MORE RESULTS IN TWO DIMENSIONS

In this section, Ω will denote a bounded open set of $\mathbb{R}^2$. For any vector field (u_1, u_2) defined in Ω, the curl $\nabla \wedge u = (D_1 u_2 - D_2 u_1)$ can be viewed as a vector perpendicular to the plane $\mathbb{R}^2 = \{(x_1, x_2)\}$. Therefore, the a priori estimates are fairly simpler and one can easily show that the curl of the solution of the Euler equation will remain bounded in the space $L^\infty(\Omega)$. For the Euler equation one can prove the existence of a weak solution for all time (with initial data in H^1), the existence and uniqueness for all time of a C^∞ (or an analytic) solution with initial C^∞ (or analytic) data.

It is possible to prove that the weak solution is the limit (when ε goes to zero) of the following Navier Stokes equation:

$$\frac{\partial u_\varepsilon}{\partial t} + u_\varepsilon \nabla u_\varepsilon - \varepsilon \Delta u_\varepsilon = -\nabla p \, , \, \nabla \cdot u_\varepsilon = 0 \; .$$

(85)

$$\nu \cdot u_\varepsilon \big|_{\partial\Omega} = 0 \, , \quad \nabla \wedge u_\varepsilon \big|_{\partial\Omega} = 0 \, , \quad u_\varepsilon(t, \cdot) = u_0(\cdot) \; .$$

One should notice that in (85) the boundary conditions have been modified, and therefore no boundary layer will appear.

Therefore the situation is rather "good" in two dimensions. However, several problems remain open. Here are two of the most important ones:

(a) We shall prove that, when the initial data is smooth, the gradient of the solution is bounded in the space $L^{\infty}(\Omega)$; more precisely, one has the estimate

(86) $$|\nabla \wedge u|_{L^{\infty}(\Omega)} \le C(\alpha) e^{\omega t} \quad (\omega = \|\nabla \wedge u_0\|_{L^{\infty}(\Omega)})$$

It seems that this estimate is the sharpest one, and it would be interesting to study the set of points $x \in \Omega$ where $\nabla u(x,t)$ is not uniformly bounded in t (for $t \to +\infty$); it may turn out to be a set of fractional Hausdorff dimensions.

(b) The boundary condition that appears in (85) is not "realistic." As noted in the introduction, it would be interesting to study the convergence of the solution of the Navier Stokes equation with the boundary condition $u_{\varepsilon}|_{\partial\Omega} = 0$, to the solution of the Euler equation, and define the set where $u_{\varepsilon}(x,t)$ converges to $u(x,t)$. Some boundary value problem for the operator studied in section II may be considered as a model (in the linear case, see for instance Bardos, Brezis and Brezis [5]).

THEOREM 6. <u>Let</u> $u_0 \in C^{1,\alpha}_{\sigma}(\Omega)$ (Ω bounded simply connected open set of $\mathbb{R}^2$ with smooth boundary $\partial\Omega$). <u>Then there exists a unique function</u> u $C(\mathbb{R}; C^{1,\alpha}_{\sigma}(\Omega))$, <u>which is a solution of the Euler equation and which is equal to</u> u_0 for $t = 0$. <u>It satisfies the a priori estimate</u>

(87) $$\|\nabla u(t,\cdot)\|_{L^{\infty}(\Omega)} \le (C/\alpha) \exp(t D\|\nabla \wedge u_0\|_{L^{\infty}(\Omega)}).$$

(In (87), the constants C and D depend only on Ω .)

REMARK 8. The hypothesis Ω is simply connected is just made for the sake of simplicity: the general case can be found in Schaeffer [37], appendix I, or Kato [21] .

REMARK 9. Following the methods of section III (which remain valid in two dimensions), it is easy to prove that the solution will be C^∞ when the initial data are C^∞ (see, for instance, Ebin & Marsden [13] or Foias, Frisch and Temam [14]), or analytic when the initial data are analytic (one may use Theorem 4 or else give direct proof, which gives a better description of the domain of analyticity (see Bardos, Benachour and Zerner [6], or Bardos and Benachour [9]).

REMARK 10. To prove the existence of a regular solution, one needs at least some estimates on the L^∞ norm of the gradient. As it will be shown, the norm of the curl is bounded in any $L^p(\Omega)$ $(1 \le p \le \infty)$; however the relation

$$\nabla \wedge u = w \ , \quad \nabla \cdot u = 0 \ , \quad u \cdot \nu|_{\partial\Omega} = 0$$

(which, when Ω is simply connected and bounded, is equivalent to the relation $-\Delta\Phi = w$, $\Phi|_{\partial\Omega} = 0$ and $u = \nabla \wedge \Phi$) does not imply that ∇u belongs to $L^\infty(\Omega)$ when w belongs to $L^\infty(\Omega)$. Therefore one needs to use the Holder spaces and prove the following lemma.

LEMMA 1. *Let* $u(x,t)$ *be a divergence free smooth vector field defined on* $\Omega\times[0,T]$, *and assume that* $u(\cdot\,,t)$ *is tangent to* $\partial\Omega$. *Let* $x(t)$ *and* $y(t)$ *be two solutions of the differential equation*:

$$(88)\qquad \dot{x}(t) = u(x(t),t)\;,\quad \dot{y}(t) = u(y(t),t)\;.$$

Let $\varphi(t) = |x(t) - y(t)|$. *Then*

$$(89)\qquad \left|\frac{\varphi(0)}{L}\right|^{\exp\left(-D\int_0^t |\nabla\wedge u|_\infty\, ds\right)} \le \frac{\varphi(t)}{L} \le \left|\frac{\varphi(0)}{L}\right|^{\left(\exp\, D\int_0^t |\nabla\wedge u|_\infty\, ds\right)}\;.$$

In (89) L is the diameter of the open set Ω and D is a constant depending only on Ω.

PROOF. Let u be the solution of the elliptic system $\nabla\cdot u = 0$, $\nabla\wedge u = w$, $u\cdot\nu|_{\partial\Omega} = 0$. Then u is given by the integral representation:

$$(90)\qquad u(x) = \int_\Omega (g(x,y)/|x-y|)\, w(y)\, dy$$

where $g(x,y)$ is a smooth function. From (90) one deduces, by a Schauder-estimate-type argument, the following relation:

$$(91)\qquad |u(x) - u(y)| \le D|x-y|\log\frac{L}{|x-y|}\,|\nabla\wedge u|_\infty\;.$$

Therefore, using (88) and (91), one obtains the differential inequality

$$(92)\qquad \left|\frac{\varphi'}{\varphi \log \frac{L}{\varphi}}\right| \le D|\nabla \wedge u|_\infty ,$$

and (89) can be deduced from (92) by an integration.

PROOF OF THEOREM 6. Taking the curl of the Euler equation one obtains in two dimensions the formula

$$(93)\qquad \frac{\partial w}{\partial t} + u \cdot \nabla w = 0$$

(because the term $w \cdot \nabla u$ is identically zero).

Therefore one may introduce the iteration scheme defined by the relations

$$(94)\qquad \begin{aligned} &\frac{\partial w^{n+1}}{\partial t} + u^n \cdot \nabla w^{n+1} = 0 , \quad w^n(0,\cdot) = \nabla \wedge u_0(\cdot) \\ &\nabla \wedge u^{n+1} = w^{n+1} , \quad \nabla \cdot u^{n+1} = 0, \quad u^{n+1} \cdot \nu|_{\partial\Omega} = 0 . \end{aligned}$$

It is clear that w^n remains constant along the characteristics (i.e., the solution of the equation

$$(95)\qquad x'(s) = u^n(x(s),s).)$$

Next we shall show that the Holder norm of w^{n+1} is bounded in some convenient way.

Let $x,y \in \Omega$, $t \ge 0$. Let one introduce the solutions $x(s)$ and $y(s)$ of (95) which satisfy the relations $x(t) = x$, $y(t) = y$.

Using Lemma 1 or the relation (92) one can show that, with

$\beta(s) = \alpha \exp(-D|\nabla \wedge u_0|_\infty \cdot s)$,

(96) $$\frac{d}{dt_+}(|w^{n+1}(x(s),s) - w^{n+1}(y(s),s)|/(\varphi(s)/L)^{\beta(s)}$$

$$\leq |w^{n+1}(x(s),s) - w^{n+1}(y(s),s)|/(\varphi(s)/L)^{\beta(s)}$$

$$\alpha e^{-D|\nabla\wedge u_0|_\infty s}\left[-D|\nabla\wedge u_0|_\infty \log\frac{L}{\varphi(s)} + \varphi'(s)/\varphi(s)\right] \leq 0 \ .$$

Therefore, one finally obtains the relation

(97) $$|(w^{n+1}(x,t) - w^{n+1}(y,t)/|x-y|^{\beta(t)}| \leq |\nabla \wedge u_0|_{0,\alpha} \ L^{(\alpha-\beta(t))} \ .$$

From the relation (97) one may deduce that, for instance, for any $T > 0$, w^n is uniformly bounded in the space $L^\infty(0,T; C_\sigma^{1,\beta(T)}(\Omega))$. Therefore one can show that w^n will converge to a limit w which belongs to the same space and which satisfies the estimate

(98) $$|w(t,\cdot)|_{0,\beta(t)} \leq L^{(\alpha-\beta(t))}|\nabla \wedge u_0|_{0,\alpha} \ .$$

The rest of the proof is easy (or at least classical); in particular, (87) is a direct consequence of (98) (compute the constants that appear in Schauder's estimate).

Finally we have the following theorem.

THEOREM 7. <u>Assume that</u> u_0 <u>satisfies the following relations</u>: $\nabla \cdot u_0 = 0$, $u_0 \cdot \nu|_{\partial\Omega} = 0$, $\nabla \wedge u_0$ $L^p(\Omega)$ $(p \geq 2)$. <u>Then there exists a function</u> u <u>belonging to the space</u> $C(\mathbb{R}_+; L^2(\Omega))$ <u>which</u>

is a solution of the Euler equation and which satisfies the relation $u(t,\cdot) = u_0$ *. Furthermore,* u *is the limit in* $C(\mathbb{R}_+; L^2(\Omega))$ *of the corresponding solution of the Navier Stokes equation with modified boundary conditions.*

Finally, $\nabla \wedge u(t,\cdot) \in L^\infty(\mathbb{R}_+; L^p(\Omega))$, *and when* $p = +\infty$ *the solution of this problem is unique.*

PROOF. It is easy to see that (85) is a well-posed problem. Multiplying by u_ε and integrating by parts one obtains:

$$(99) \qquad \tfrac{1}{2}\int_\Omega |u_\varepsilon(t,x)|^2\,dx + \varepsilon\int_0^T\int_\Omega |\nabla u_\varepsilon(t,x)|^2\,dx = \tfrac{1}{2}\int_\Omega |u_0(x)|^2\,dx\,,$$

which is the classical energy estimate. Next one takes the curl of the first equation of (85) and obtains the following expression (here there appears the fact that the dimension is two).

$$(100) \qquad \frac{\partial}{\partial t}(\nabla \wedge u_\varepsilon) - \varepsilon\,\Delta(\nabla \wedge u_\varepsilon) + u_\varepsilon \cdot \nabla(\nabla \wedge u_\varepsilon) = 0\ .$$

Due to the modified boundary condition one obtains $\nabla \wedge u_\varepsilon|_{\partial\Omega} = 0$ and all the a priori estimates are easy. The relation $\nabla \wedge u \in L^\infty(\mathbb{R}_+; L^p(\Omega))$ implies the relation $u \in L^\infty(\mathbb{R}_+; W^{1,p}(\Omega))$ except for $p = +\infty$. Then one has the following estimate:

$$(101) \qquad |u(t,\cdot)|_{W^{1,p}(\Omega)} \le C\,p\,|\nabla \wedge u(t,\cdot)|_{L^\infty(\Omega)}\ .$$

This estimate is used to prove the uniqueness of the weak solution when $\nabla \wedge u_0 \in L^{\infty}(\Omega)$. The proof is omitted here (see Bardos [4] or Youdovich [41] for details).

V. SOME CONCLUDING REMARKS

As long as a regular solution exists (for finite time in three dimensions, for all time in two dimensions), the situation is rather clear. The next step would be to obtain an estimate in three dimensions for all time (or in two dimensions for a domain with boundary $\partial\Omega$ and boundary condition $u_\varepsilon|_{\partial\Omega} = 0$) on the regularity of the solution which should be independent of time and of the viscosity parameter ε , as is for instance proved for the Burger-Hopf equation ($\hat{u}_\varepsilon(t,k) \sim \frac{C}{k}$, $u_\varepsilon(t,x)$ is bounded uniformly in BV -space of bounded variation functions-). The proof of an estimate of that type may rely on the ideas of Kolmogorov (1941). (See also Frisch, Sulem and Nelkin [15] for recent developments and extensive bibliography.) $u_\varepsilon(x,t)$ will denote the solution of the Navier Stokes equation with viscosity $\varepsilon > 0$. We shall consider a discrete sequence of the scale of eddies $\ell_n = \ell_0 2^{-n}$ and of wave numbers $k_n = \ell_n^{-1}$. The kinetic energy per unit mass in the scale $\sim\ell_n$ is defined as

$$E^{\varepsilon}_n(t) = \int_{k_n}^{k_{n+1}} E_\varepsilon(k,t)\, dk$$

where $E_\varepsilon(k,t)$ is the kinetic energy per unit mass and unit wave number. Using the Parseval relation, we may write

$$u_{\epsilon n}^2(t) = E_n^\epsilon(t) \quad .$$

Next we define a sequence t_n of time (eddy turnover time) which will satisfy the relation

$$t_n = \ell_n / u_{\epsilon n}(t_n) \quad .$$

This sequence will be the characteristic rate at which excitation of the scale ℓ_n is fed into scales ℓ_{n+1} .

The rate of transfer of energy per unit mass from n eddies to $(n+1)$ eddies is then given by

$$(102) \qquad \varepsilon_n \sim E_n/t_n \sim v_n^3/\ell_n \qquad (E_n = E_n^\varepsilon(t_n) \ , \quad v_n = u_{\epsilon n}(t_n)\ ) \ .$$

Since the global energy dissipation is bounded, one may assume that $\epsilon_n = \bar{\varepsilon}$ is constant (independent of ε , t and n). Finally, by solving the equations, we obtain

$$(103) \qquad v_n \sim (\bar{\varepsilon}\ell)^{1/3} \ , \quad E_n \sim (\bar{\varepsilon}\ell_n)^{2/3}$$

which, by the Fourier transformation, yields the spectrum

$$E(k) \sim (\bar{\varepsilon})^{2/3} k^{-2/3} \quad .$$

Of course, this argument is <u>not a proof</u>. However, it suggests that the Fourier transformation of the solution should be uniformly bounded in $H^{2/3}(\mathbb{R}^n)$.

The next problem is the regularity for all time and small viscosity of the solution of the Navier Stokes equation. It is known that the solution of the Navier Stokes equation is analytic for all time when the initial data are small enough with respect to the viscosity (see, for instance, Kato and Fujita [23]). On the other hand, the following equation (with modified dissipativity) has been considered by Lions [30]:

$$(104) \qquad \frac{\partial u}{\partial t} + (-\Delta)^{\alpha} u + u \nabla u = -\nabla p \ , \quad \nabla \cdot u = 0 \ , \quad \text{in} \ \mathbb{R}^3 \ .$$

It has been proved that, for $\alpha \geq 5/4$, the solution is always regular. The question is then the following: is the operator $(-\Delta)$ strong enough to counterbalance the effect of the non-linear term? The two following remarks are interesting in this regard:

REMARK 11. The additional hypothesis that u is bounded in $H^{2/3}(\mathbb{R}^n)$, which can be deduced from the Kolmogorov law, is not enough to ensure the regularity of the solution of the Navier Stokes equation.

REMARK 12. It is possible to construct models of parabolic non-linear equation of the type (104) with modified dissipativity that exhibit singularities after a finite time. An example is studied in Bardos, Frisch, Penel and Sulem [8]. Another one can be constructed with the Burger equation:

$$(105)\qquad \frac{\partial u_\varepsilon}{\partial t} + \varepsilon\left(-\frac{\partial^2}{\partial x^2}\right)^\alpha u_\varepsilon + u_\varepsilon^\alpha \frac{\partial u_\varepsilon}{\partial x} = 0 \ . \qquad 0 < \alpha < \tfrac{1}{2} \ .$$

It is easy to see that

$$(106)\qquad |u_\varepsilon(t,k)| \le (1/|k|)\int_{-\infty}^{+\infty} \left|\frac{\partial}{\partial x} u(0,x)\right| dx \ .$$

Using (106) one can show via Parseval the following relation (use the relation $0 < \alpha < \frac{1}{2}$).

$$(107)\qquad \lim_{\varepsilon\to 0} \int_0^t \int_{-\infty}^{\infty} \varepsilon\left(-\frac{\partial^2}{\partial x^2}\right)^\alpha u_\varepsilon \cdot u_\varepsilon \, dx = 0$$

for any t such that the solution is regular on the interval $[0,t]$.

It is clear that, when ε goes to zero, u_ε goes to u , which is the solution of the Burger equation studied in section II. Now if (107) were true for any $t > 0$, then it would be possible to prove that for any $t \ge 0$,

$$(108)\qquad \int_{-\infty}^{+\infty} |u(x,t)|^2 dx = \int_{-\infty}^{+\infty} |u(x,0)|^2 \, dx \ .$$

Morever, (108) is false when the solution of the Burger equation becomes singular. Therefore, for some $\varepsilon > 0$ and some $t > 0$, the solution $u_\varepsilon(x,t)$ will not be regular.

REFERENCES

1. AGMON, S., DOUGLIS, A. and NIRENBERG, L., Estimates near the boundary for solutions of elliptic partial differential equations satisfying general boundary conditions. Comm. Pure Appl. Math. 12(1959), 623-727.

2. BAOUENDI, M.S. and GOULAOUIC, C., Problemes de Cauchy pseudo differentiels analytiques. Seminaire Goulaouic-Schwartz, 1975-76.

3. BARDOS, C., Analycité de la solution de l'équation d'Euler dans un ouvert de $\mathbb{R}^n$. Note C.R. Acad. Sc., Paris A. 283(1976), 255-258.

4. BARDOS, C., Existence et unicité de la solution de l'équation d'Euler en dimension deux. J. Math. Anal. Appl. 40(1972), 769-790.

5. BARDOS, C., BREZIS, D. and BREZIS, H., Perturbations singulieres et prolongements maximaux d'opérateurs positifs. Arch. Rat. Mech Anal. 53(1973-1974), 69-100.

6. BARDOS, C., BENACHOUR, S. and ZERNER, M., Analycité des solutions periodiques de l'équation d'Euler en deux dimensions. Note C.R. Acad. Sc., Paris A. 282(1976), 995-998.

7. BARDOS, C. and FRISCH, U., Finite time regularity for bounded and unbounded ideal incompressible fluid using Hölder estimates. Turbulence and Navier Stokes equation, Orsay (1975). Lecture Notes in Mathematics (565), Springer-Verlag, 1-14.

8. BARDOS, C., FRISCH, U., PENEL, P., and SULEM, P.L., Modified Dissipativity for a non linear evolution equation arising in Turbulence. Turbulence and Navier Stokes equation., Orsay (1975). Lecture Notes in Mathematics (565), Springer-Verlag, 14-23. (A detailed version will appear in the Arch. Rat. Mech. Anal.)

9. BARDOS, C., and BENACHOUR, S., Domaine d'Analycité des solutions de l'equation d'Euler dans un ouvert de $\mathbb{R}^n$. (To appear in the Annali di Pisa.)

10. BENACHOUR, S., Analycité des solutions de l'équation d'Euler (thesis, Nice, June 1976; submitted to the Arch. Rat. Mech. Anal.)

11. BOURGUIGNON, J.P. and BREZIS, H., Remarks on the Euler equation. J. Functional Analysis 15(1974), 341-363.

12. CRANDALL, M.G., The semigroup approach to first order quasilinear equations in several space variables. Israel J. of Math. 12 (1972), 108-132.

13. EBIN, D. and MARSDEN, J., Groups of Diffeomorphism and the motion of an incompressible fluid. Ann. of Math. 92(1970), 102-163.

14. FOIAS, C., FRISCH, U. and TEMAM, R., Existence des solutions C^∞ des equations d'Euler. C.R. Acad. Sc., Paris A. 280 (1975), 505-508.

15. FRISCH, U., NELKIN, M. and SULEM, P.L., A simple dynamical model of intermittent fully developed turbulence. Submitted to J. Fluid Mech.

16. FROSTMAN, O., Potentiel d'Equilibre et capacité des ensembles. Thesis, Lund (1935).

17. HOLDER, E., Uber die Unberchankle Fortsezbarkei einer Stetigene ebener Bewegung in einer unbegrenzten-inkompressiblen Flüssigkeit. Math. Z. 37(1933), 727-732.

18. HOPF, E., The partial differential equation $u_t + uu_x = \mu u_{xx}$. Comm. Pure Appl. Math. 3(1960), 201-250.

19. KAHANE, C., Solutions of Mildly singular integral equations Comm. Pure Appl. Math. 18(1965), 593-626.

20. KAHANE, J.P., Mesures et dimension. Turbulence and Navier Stokes , Orsay (1975). Lecture Notes in Mathematics (565), Springer-Verlag, 94-104.

21. KATO, T., On the classical solution of the two-dimensional non-stationary Euler equation. Arch. Rat. Mech. Anal. 25 (1967), 302-324.

22. KATO, T., Non-stationary flows of viscous and ideal fluids in $\mathbb{R}^3$. J. Funct. Anal. (1972), 296-305.

23. KATO, T. and FUJITA, M., On the non-stationary Navier Stokes system. Rend. Sem. Mat. Univ. Padova 32(1962), 243-260.

24. KRUCKOV, N.S., First order quasilinear equations in several space independent variables. Math. Sbornik. 81(123) (1970), Math. U.S.S.R. Sbornik 10 (1970), 217-243.

25. LADYZENSKAIA, O.A.and URALTCEVA, N., The mathematical theory of viscous incompressible flow. Gordon and Breach, New York (1969).

26. LAX, P.D., Hyperbolic systems of conservation laws. Comm. Pure Appl. Math. 10(1957), 537-566.

27. LAX, P.D., Shock waves and entropy in non-linear functional analysis. Proc. Symposium Univ. of Wisconsin (1971). (Edited by Zarantonello.)

28. LERAY, J., Etudes de diverses equations non lineaires et de quelques problemes que pose l'hydrodynamique. J. Math. Pure Appl. 9(12) (1933), 1-82.

29. LIONS, J.L., Problemes aux limites variationnels. Cours Université de Montreal (Summer, 1961).

30. LIONS, J.L., Quelques methodes de resolution des problèmes aux limites non linéaires. Dunod, Paris (1969).

31. LICHTENSTEIN, I., Uber einige problem der Hydrodynamic. Math. Z. 23(1925).

32. MORREY, C.B., Multiple integral in the calculus of variations. Springer-Verlag 130 (1966).

33. NIRENBERG, L., An abstract form of the non-linear Cauchy-Kowalewski Theorem. J. Diff. Geometry 6, 4 (1972), 561-576.

34. NISHIDA, T., To appear in the Journal of Differential Equations.

35. OLEINIK, O., Uspeki Math Nauk 14, 2 (86), 165-170.

36. OVSJANNIKOV, L.V., Singular Operators in Banach Scales. Dokl. Akad. Nauk. U.S.S.R. 163(1965), 819-822.

37. SCHAEFFER, A.C., Existence Theorem for the flow of an incompressible fluid in two dimensions. Trans. of the A.M.S. 42(1937), 497-513.

38. TEMAM, R., On the Euler equation of incompressible perfect fluids. J. Funct. Anal. 20(1975), 32-49.

39. VOLPERT, A.J., The space B.V. and quasi-linear equations. Math. Sbornik. 73(115) (1967), 255-302. Math. U.S.S.R. Sbornik 2 (1967), 225-267.

40. WOLIBNER, W., Un theoreme sur l'existence du mouvement plan d'un fluid parfait, homogene et incompressible pendant un temps infiniment long. Math. 2, 37 (1933), 727-738.

41. YOUDOVICH, V.I., Ecoulement non-stationnaire d'un fluide ideal non visqueux. J. Math. Numer. Phys. Math. 6(1965), 1032-1066.

ASYMPTOTIC BEHAVIOR OF A MODEL IN POPULATION GENETICS

H. F. Weinberger*

School of Mathematics
University of Minnesota

1. INTRODUCTION

A model for the spatial spread of an advantageous gene in a population living in a homogeneous one-dimensional habitat was proposed by R.A. Fisher [6]. In this model the time evolution of the fraction $u(x,t)$ of the advantageous gene in the population at the point x at time t is governed by a partial differential equation of the form

$$\frac{\partial u}{\partial t} = \frac{\partial^2 u}{\partial x^2} + f(u) \quad . \tag{1.1}$$

Fisher observed that for his equation, in which $f(u) = u(1-u)$, there are travelling waves of all speeds greater than or equal to 2, and conjectured that a new mutant gene would be propagated with the asymptotic speed 2. The idea that a nonlinearity can cause a parabolic equation to have travelling wave solutions and a finite asymptotic speed of propagation is mathematically intriguing.

---

* This work has been supported by the National Science Foundation through Grants NSF GP 37660X and NSF MCS76-06128.

Kolmogoroff, Petrovsky, and Piscounoff [11] proved that when at time $u = 1$ for $x < 0$ and $u = 0$ for $x > 0$, then the solution $u(x,t)$ approaches, in a certain sense, the slowest travelling wave, provided $f(0) = f(1) = 0$, $f'(0) > 0$, and $f''(u) < 0$.

The results of Fisher and of Kolmogoroff, Petrovsky, and Piscounoff were extended to more general functions $f(u)$ and to more general initial conditions by Kanel' [9, 10].

More recently, it was shown by Aronson and Weinberger [1] that for a large class of equations of the form (1.1) there is a slowest wave speed c^*, and that if at $t = 0$ u vanishes outside a bounded set, then u becomes 1 (that is, the advantageous gene takes over) at an asymptotic speed c^*. These results were also shown to hold in more than one dimension when the operator $\partial^2/\partial x^2$ in (1.1) is replaced by the Laplace operator [2]. This is important, since habitats tend to have either two or three dimensions.

For a special class of functions $f(u)$, Fife and McLeod [4] have proved the more precise result that a solution of the one-dimensional equation with initial values of bounded support converges to a sum of two waves travelling in the two outward directions.

As was already discussed in [1], the Fisher model is based on some assumptions of doubtful validity. This paper presents a model in which time occurs in discrete steps designed to simulate synchronous generations. The model is derived in Section 2 under more (though still not entirely) realistic hypotheses about the life cycle. Our mathematical model is described by a recursion formula

$$u_{n+1} = Q[u_n]$$

which links the state u_n of the nth generation to the state u_{n+1} of the $(n+1)$st generation. The function $u_n(x)$ represents the fraction of the advantageous gene present among the genes in those individuals of the nth generation which are born near x. Q is an operator acting on the state space U, which is the set of functions with values on the interval $[0,1]$. The homogeneity and isotropy of the habitat is reflected in the fact that Q is invariant under translations and rotations.

A mathematical model is, of course, useful only if its solutions can be analyzed more easily than the system which is being modelled. Consequently, the remainder of this paper is devoted to showing how useful information about solutions of the recursion $u_{n+1} = Q[u_n]$ can be obtained. For the sake of exposition, we shall only treat the simplest cases and indicate some other cases which can be treated.

We shall obtain all the qualitative features of the Fisher model. We define a speed c^* by means of (3.6). In Section 3 we state three theorems for the heterozygote intermediate case which show that c^* is the asymptotic speed of propagation for solutions of the recursion formula. These results are proved in the next two sections. Section 6 shows how to extend these results to the heterozygote superior case.

In Section 7 we show that there are travelling waves with speed c precisely when $|c| \geq c^*$, and we find the asymptotic behavior of the waves. Section 8 shows how the Fisher model can formally be derived from our model of Section 2. More importantly, we show how the study of the Fisher model and an important generalization of it can be reduced to the study of recursions of the form $u_{n+1} = Q[u_n]$.

There are connections between our model and a probabilistic model of Hammersley [8]. We discuss these connections briefly at the end of Section 2.

The model presented in Section 2 is, of course, rather crude. It can be refined by allowing several periods of selection, reaping, and migration to occur during a single life cycle. This leads to a recursion formula of the form $u_{n+1} = R[u_n]$ where R is the composition of several operators of the form of the operator Q which is studied here. One can easily extend our results to such a composition of operators.

As is indicated in the discussion at the end of Section 8, many of the simplifying assumptions which will be made here are not needed, and more general results of the kind presented here will be published elsewhere.

Our model is easily extended to encompass a habitat which is not homogeneous. One again obtains a recursion $u_{n+1} = Q[u_n]$, but Q is no longer translation invariant. One could then hope to find analogues of the results of Fife and Pelletier [5], Fleming [7], Nagylaki [12], Slatkin [13], and others on clines for the Fisher equation with $f(u)$ replaced by $f(u,x)$.

Finally, we mention that the usefulness of the recursion model is not confined to population genetics. Equations of the form $u_{n+1} = Q[u_n]$ arise, for instance, in connection with the study of population growth of a migrating species, where u_n represents the population density.

Many of the ideas presented here were developed in discussions with D.G. Aronson, and I am grateful for his valuable suggestions and criticisms.

2. A SIMPLE MODEL

The Fisher model is intended to describe the following biological situation. One gene of a certain species occurs in two variant forms, which are called alleles and denoted by A and a . Each individual of the species has two of these alleles at the chromosome locus corresponding to the gene in question, each of which may be of type A or a . Thus, with respect to the particular gene under discussion, there are three genotypes, which we denote by AA , Aa , and aa . Individuals of the genotypes AA and aa are called homozygotes, while Aa individuals are called heterozygotes.

We assume that the populations of the three genotypes are continuously distributed with the population densities $\rho_{AA}(x)$, $\rho_{Aa}(x)$, and $\rho_{aa}(x)$ in a habitat which is a Euclidean space of $N(=1, 2, \text{ or } 3)$ dimensions.

The population densities vary with time. In particular, mating occurs, at random and without regard to genotype, among nearby individuals, the individuals migrate, and, most importantly, selection occurs, in the sense that individuals of the three genotypes have slightly different survival rates which depend only on the genotype with respect to the gene under consideration.

It is clear that this set of assumptions should lead to a set of three equations in the three unknowns ρ_{AA} , ρ_{Aa} , and ρ_{aa} , rather than to a single equation like Fisher's equation in the variable

$$u = \frac{2\rho_{AA} + \rho_{Aa}}{2(\rho_{AA} + \rho_{Aa} + \rho_{aa})} . \tag{2.1}$$

The three equations are studied in [1], where it is shown that one can never obtain the Fisher equation exactly.

What, then, must one assume to obtain a single equation in the single variable u ? Clearly, one needs two relations between the three variables ρ_{AA} , ρ_{Aa} , and ρ_{aa} . These two relations, which are implicit in Fisher's derivation, are

(i) <u>The Hardy-Weinberg law</u>:

$$\rho_{AA} : \rho_{Aa} : \rho_{aa} = u^2 : 2u(1-u) : (1-u)^2 . \tag{2.2}$$

(ii) <u>The total population density</u> $\rho_{AA} + \rho_{Aa} + \rho_{aa}$ <u>is independent of</u> x .

The Hardy-Weinberg law holds automatically for the offspring produced by a set of random matings. However, such a mating does not change the ratios $\rho_{AA} : \rho_{Aa} : \rho_{aa}$ of the parents, so that if these ratios have been altered through selection, the Hardy-Weinberg law will, in general, be satisfied at the time of the birth of a new generation only if the entire parent generation dies or is somehow removed from consideration when the new generation is born. This property describes, for example, the behavior of annual flowers or of various species of insects, in which the generations are synchronized so that births occur at discrete times at which the preceding generation dies. It cannot hold in a species in which mating occurs continuously, which is the situation that is supposedly described by the Fisher model.

The hypothesis (2.2 ii) of constant total population density is implicit in the use by Fisher of the diffusion term to model migration. One must only observe that while individuals may migrate, fractions do not. In order to make u proportional to a population density, one assumes that the denominator in (2.1) is constant whenever migration is taking place.

In order to obtain a deterministic model for what is, in fact, a series of stochastic processes, one always needs to make the additional assumption that *the outcome of any stochastic process is exactly its expected value*. That is, we assume that all variances are zero.

The simplest model incorporating this last hypothesis in which the hypotheses (2.2) are satisfied is described by the following scheme:

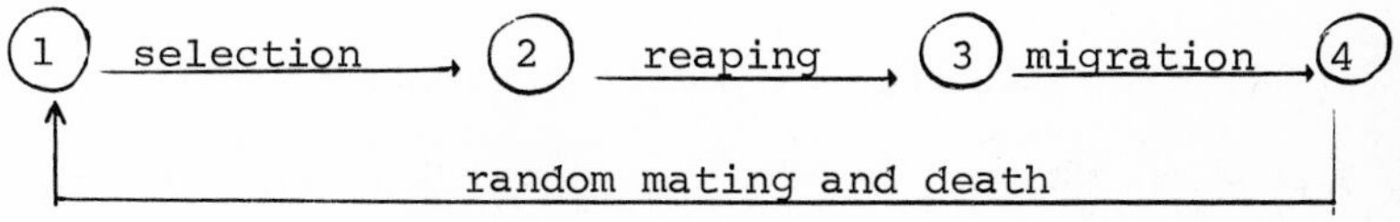

Stage 1: A new generation has just been born and the old generation has died. $u(x)$ is the fraction of the allele A in the parent population at x, as defined by (2.1), and if the total population density of the old population was a constant μ, then the population densities of the new generation are

$$\rho_{AA}(x) = \varphi\mu u(x)^2 ,$$

$$\rho_{Aa}(x) = 2\varphi\mu u(x)(1-u(x)) ,$$

$$\rho_{aa}(x) = \varphi\mu(1-u(x))^2 .$$

The constant φ represents the number of offspring per individual of the parent generation.

A selection process with the different survival rates α for AA individuals, β for Aa individuals, and γ for aa individuals leads to Stage 2. The population densities have become

$$\rho_{AA} = \alpha\varphi\mu u^2 ,$$

$$\rho_{Aa} = 2\beta\varphi\mu u(1-u) ,$$

$$\rho_{aa} = \gamma\varphi\mu (1-u)^2 .$$

During the next phase of development the total population density is reduced to the constant carrying capacity μ of the habitat by a random process which affects all the genotypes equally. For example, the selection process may occur in the larval stage of an insect, and μ may represent the density of sites on which pupae can be formed. We called this process reaping. At the end of the reaping process the population arrives at Stage 3. The population densities at this stage are

$$\rho_{AA} = \mu \frac{\alpha u^2}{\alpha u^2 + 2\beta u(1-u) + \gamma(1-u)^2} ,$$

$$\rho_{Aa} = \mu \frac{2\beta u(1-u)}{\alpha u^2 + 2\beta u(1-u) + \gamma(1-u)^2} ,$$

$$\rho_{aa} = \mu \frac{\gamma(1-u)^2}{\alpha u^2 + 2\beta u(1-u) + \gamma(1-u)^2} .$$

The population now undergoes a random migration. We assume that a fraction $k(y)dy$ of the population originally at x winds up in the interval $(x_1+y_1, x_1+y_1+dy_1) \times (x_2+y_2, x_2+y_2+dy_2) \times \ldots \times (x_N+y_N, x_N+y_N+dy_N)$. Of course, $k(y) \geq 0$, and since every individual goes somewhere,

$$\int_{R^N} k(y)\,dy = 1 \quad .$$

That is, k is a probability kernel. The fact that $k(y)$ is independent of x means that the migration process is homogeneous (that is, translation invariant).

After the migration the population is at Stage 4 with population densities

$$\rho_{AA}(x) = \mu \int_{R^N} k(x-y) \frac{\alpha u(y)^2}{\alpha u(y)^2 + 2\beta u(y)(1-u(y)) + \gamma(1-u(y))^2}\, dy \quad ,$$

$$\rho_{Aa}(x) = \mu \int_{R^N} k(x-y) \frac{2\beta u(y)(1-u(y))}{\alpha u(y)^2 + 2\beta u(y)(1-u(y)) + \gamma(1-u(y))^2}\, dy \quad ,$$

$$\rho_{aa}(x) = \mu \int_{R^N} k(x-y) \frac{\gamma(1-u(y))^2}{\alpha u(y)^2 + 2\beta u(y)(1-u(y)) + \gamma(1-u(y))^2}\, dy \quad .$$

The total population density $\rho_{AA} + \rho_{Aa} + \rho_{aa}$ is the constant μ, as it was in the parent population just before it mated and died. The function $u(x)$ defined by (2.1), however, has now become

$$\int_{R^N} k(x-y) \frac{\alpha u(y)^2 + \beta u(y)(1-u(y))}{\alpha u(y)^2 + 2\beta(y)(1-u(y)) + \gamma(1-u(y))^2}\, dx \quad . \qquad (2.3)$$

The population now undergoes random mating and dies, thus completing the life cycle at Stage 1. We let $u_n(x)$ be the function defined by (2.1) at Stage 1 of the n th generation. Then $u_{n+1}(x)$ is given by (2.3) with u replaced by u_n .

We define the function of one variable

$$g(u) \equiv \frac{\alpha u^2 + \beta u(1-u)}{\alpha u^2 + 2\beta(1-u) + \gamma(1-u)^2} , \tag{2.4}$$

and the operator

$$Q[u](x) \equiv \int_{R^N} k(x-y)g(u(y))dy \quad . \tag{2.5}$$

The evolution of the function $u_n(x)$ in time for our model then has the form

$$u_{n+1} = Q[u_n] \quad . \tag{2.6}$$

Since α , β , and γ are positive, the function $g(u)$ clearly has the properties

$$\begin{aligned} g(0) &= 0 \\ g(1) &= 1 \\ g'(u) &\geq 0 \quad \text{for} \quad 0 < u < 1 \quad . \end{aligned} \tag{2.7}$$

In particular $0 \leq u \leq 1$ implies $0 \leq g(u) \leq 1$ and hence, since k is a probability kernel, $0 \leq Q[u] \leq 1$. Thus Q maps the set U of measurable functions $u(x)$ with values in $[0,1]$ into itself. It follows from the definition (2.1) that we are only interested in functions u_n in this set. In addition, Q is translation invariant. That is, for any constant displacement b

$$Q[u(x-b)](z) = Q[u(x)](z-b) \quad . \tag{2.8}$$

We are interested in the behavior as n approaches infinity of solutions u_n of the recursion (2.6). This behavior depends only on the selection ratios $\alpha:\beta:\gamma$. Since interchanging the names of the alleles A and a merely interchanges α and γ , we shall assume without loss of generality that the names are chosen so that $\alpha \geq \gamma$. That is, the survival rate of the heterozygote AA is no less than that of the heterozygote aa . There are then three distinct cases:

(i) heterozygote intermediate: $\alpha \geq \beta \geq \gamma$, $\alpha > \gamma > 0$

(ii) heterozygote superior: $\beta > \alpha \geq \gamma > 0$ (2.9)

(iii) heterozygote inferior: $\alpha \geq \gamma > \beta > 0$.

We have not as yet been able to treat the heterozygote inferior case or the special case $\alpha > \beta = \gamma$ of (i) . In this paper we shall only treat the heterozygote superior case and the subcase

$$\alpha \geq \beta \geq \tfrac{1}{4}[\gamma + (\gamma^2 + 8\alpha\gamma)^{\frac{1}{2}}] \ , \quad a > \gamma \tag{2.10}$$

of the heterozygote intermediate case. The inequality (2.10) or (2.9 ii) implies that $g(u)$ is bounded by its linearization about $u = 0$:

$$g(u) \leq g'(0)u \ .$$

For the sake of simplicity we shall also make some assumptions about the migration kernel k . We assume that the habitat is isotropic (invariant under rotation) so that k depends only on the distance of migration:

$$k = k(|y|) \ . \tag{2.11}$$

We also assume that an individual can migrate no farther than a certain distance B in its lifetime, so that

$$k(|y|) = 0 \quad \text{for} \quad |y| > B \ . \tag{2.12}$$

Finally, we assume that $k(|y|)$ is continuous. Our results can actually be proved under considerably weaker hypotheses. For instance, all our results except for the asymptotic estimate (7.5) remain valid if (2.12) is replaced by the hypothesis that $e^{\mu|y|}k(|y|)$ has a finite integral for all real μ. Moreover, most of the results remain valid if the inequality (2.10) is replaced by the weaker inequality (2.9 i) together with $\beta > \gamma$. However, the asymptotic speed c^* may no longer be computed from (3.6) in this case.

We could as well have worked with the variable $v_n = g(u_n)$, which represents the fraction (2.1) at Stage 3 rather than Stage 1 of the life cycle. By taking g of both sides of (2.6), one finds that v_n satisfies the recursion

$$v_{n+1}(x) = g(\int k(|x-y|)\, v_n(y)\, dy) \quad . \tag{2.13}$$

Since $g(0) = 0$ and $g(1) = 1$, all our theorems apply immediately to the sequence $v_n(x)$.

The special one-dimensional initial value problem for the recursion (2.13) with

$$v_0(x) = \begin{cases} 1 & \text{for } x < 0 \\ 0 & \text{for } x > 0 \end{cases}$$

has been considered in another context by Hammersley [8]. He obtained a formula equivalent to (3.6) for the asymptotic speed, and showed that $v_n(nc)$ approaches 1 for $c < c^*$ and 0 for $c > c^*$.

3. LARGE-TIME BEHAVIOR OF A HETEROZYGOTE INTERMEDIATE POPULATION

In this section we shall state some results about the behavior for large n of the solutions of the recursion

$$u_{n+1} = Q[u_n] \tag{3.1}$$

where Q is an operator of the form

$$Q[u](x) = \int k(|x-y|)g(u(y))dy$$
$$= \int k(|z|)g(u(x-z))dz \ .$$

(The second form comes from the first by the simple change of variables $z = x - y$, and we shall use the two forms interchangeably.) Unless otherwise specified, all integrals will be understood to extend over the Euclidean N-space R^N.

We shall assume throughout this article that $k(|x|)$ is a continuous nonnegative function with

$$\int k(|x|)\,dx = 1 \ ,$$

and that

$$k(|x|) = 0 \quad \text{for} \quad |x| > B \ . \tag{3.2}$$

The function $g(u)$ need not be of the form (2.4) but we shall assume that it is continuously differentiable on the interval $[0,1]$ and that it has the properties

$$g(0) = 0 \ , \quad g(1) = 1 \ , \quad g'(u) \geq 0 \quad \text{in} \quad [0,1] \ , \tag{3.3}$$

$$g(u) > u \quad \text{in} \quad (0,1) \ , \tag{3.4}$$

and

$$g(u) \leq g'(0)u \quad \text{in} \quad [0,1] \ . \tag{3.5}$$

When g is of the form (2.4) with α, β, and γ positive, (3.3) is true, (3.4) is equivalent to the inequality (2.9 i) which characterizes the heterozygote intermediate case, and (3.5) follows from the stronger inequality (2.10).

Since the integral of k is one, the condition (3.3) implies that Q takes the set U of measurable functions in R^N with values in $[0,1]$ into itself, and we shall work on this set.

We define the constant

$$c^* = \min_{\mu>0} \left\{ \frac{1}{\mu} \log \left[g'(0) \int e^{\mu x_1} k(|x|)\,dx \right] \right\} . \qquad (3.6)$$

It will be shown in the next section that this minimum exists. The following results state that c^* is an asymptotic propagation speed for the spread of the advantageous allele A.

THEOREM 1. *Let* u_n *be a solution of the recursion* (3.1), *let* $0 \le u_0(x) \le 1$, *and suppose that* u_0 *has bounded support*:

$$u_0(x) = 0 \quad \text{for} \quad |x| > b . \qquad (3.7)$$

Then

$$\lim_{n\to\infty} \max_{|x| \ge nc^*} u_n(x) = 0 . \qquad (3.8)$$

THEOREM 2. *Let* u_n *be a solution of the recursion* (3.1), *let* $0 \le u_0 \le 1$, *and suppose that* u_0 *is not identically zero in the sense that its integral is positive. Then for each* $c < c^*$

$$\lim_{n\to\infty} \min_{|x| \le nc} u_n(x) = 1 . \qquad (3.9)$$

In addition, we shall prove the following result, which shows that the equilibrium state $u_n \equiv 1$, unlike the state $u_n \equiv 0$, is stable.

THEOREM 3. *Let* u_n *be a solution of the recursion* (3.1) *with* $0 \le u_0 \le 1$, *and suppose that* $u_0(x)$ *is bounded away from* 0 *outside a bounded set*:

$$u_0(x) \ge \rho_0 > 0 \quad \text{for} \quad |x| > b \; . \tag{3.10}$$

Then u_n *converges to* 1 *uniformly in* x *as* $n \to \infty$.

We remark that the statements of Theorems 2 and 3 and that of Theorem 1 with c^* replaced by any $c > c^*$ remain valid when the condition (3.5) is replaced by the condition $g'(0) > 1$. If g is of the form (2.4), this condition is satisfied when $\alpha \ge \beta \ge \gamma$. However, in this case c^* may no longer be given by the formula (3.6). The proofs of these results will be published elsewhere.

All our proofs will be based on the following comparison result.

PROPOSITION 1. *Let* R *be an operator which takes an ordered set* S *of functions into itself and which has the monotonicity property*

$$v \le w\,, \quad v, w \in S \Rightarrow R[v] \le R[w]\,. \tag{3.11}$$

If the sequences v_n *and* w_n *of elements of* S *satisfy the inequalities*

$$\begin{aligned} v_{n+1} &\le R[v_n]\,, \\ w_{n+1} &\ge R[w_n] \qquad n = 0,1,2 \ldots \end{aligned} \tag{3.12}$$

<u>and if</u>

$$v_0 \le w_0 \quad ,$$

<u>then</u>

$$v_n \le w_n$$

<u>for</u> $n \ge 0$.

PROOF. The proof is by induction. If $v_n \le w_n$, then by (3.12) and (3.11)

$$v_{n+1} \le R[v_n] \le R[w_n] \le w_{n+1} \quad .$$

Since $v_0 \le w_0$, the result is proved.

4. PROOF OF THEOREM 1

We shall begin by showing that the constant c^* is well-defined by (3.6). We define the two functions

$$\Phi(\mu) = \frac{1}{\mu} \log\left\{ g'(0) \int e^{\mu x_1} k(|x|)dx \right\} , \quad \text{for} \quad \mu > 0 \quad , \tag{4.1}$$

and

$$\psi(\mu) = \frac{\int x_1 e^{\mu x_1} k(|x|)dx}{\int e^{\mu x_1} k(|x|)\, dx} \quad . \tag{4.2}$$

Differentiation and a simple rearrangement show that

$$\psi'(\mu) = \frac{\int [x_1 - \psi(\mu)]^2 e^{\mu x_1} k(|x|)dx}{\int e^{\mu x_1} k(|x|)dx} > 0 \quad . \tag{4.3}$$

We shall establish the following lemma.

LEMMA 4.1. <u>The function $\Phi(\mu)$ attains its minimum value c^* at a unique positive value μ^* of μ, and</u>

$$\Psi(\mu^*) = \Phi(\mu^*) = c^* \quad . \tag{4.4}$$

<u>Moreover, Φ is decreasing for $\mu < \mu^*$</u> .

PROOF. Differentiation shows that

$$\Phi'(\mu) = -\frac{1}{\mu}[\Phi(\mu) - \Psi(\mu)] \quad , \tag{4.5}$$

and hence that

$$(\mu^2\Phi')' = \mu\Psi' \quad . \tag{4.6}$$

We therefore see from (4.3) that $\mu^2\Phi'$, like Ψ, is increasing. Consequently Φ can have no local maximum. It has at most one local minimum μ^*, to the left of which Φ decreases and to the right of which it increases.

It remains to be shown that such a minimum exists. The inequalities (3.4) and (3.5) imply that

$$g'(0) > 1 \quad . \tag{4.7}$$

Since the integral of k is 1, we see from the definition (4.1) that Φ approaches $+\infty$ as μ decreases to 0.

We suppose that B is the smallest constant for which (3.2) is true. Then $k(\xi)$ is positive on the interval $(B-\delta, B)$ provided

δ is positive and sufficiently small. Clearly

$$\Phi(\mu) \geq \frac{1}{\mu} \log\left\{g'(0) \int_{x_1 \geq B-\delta} e^{\mu x_1} k(|x|)dx\right\} \tag{4.8}$$

$$\geq B - \delta + \frac{1}{\mu} \log\left\{g'(0) \int_{x_1 \geq B-\delta} k(|x|)dx\right\} .$$

We choose δ so small that

$$\int_{x_1 \geq B-\delta} k(|x|)dx \leq \frac{1}{2g'(0)} .$$

Then since $k = 0$ for $|x| \geq B$ and the integral of k is 1,

$$\Phi(\mu) \leq \frac{1}{\mu} \log\left\{g'(0) \int_{x_1 \leq B-\delta} e^{\mu x_1} k(|x|)dx + g'(0) \int_{x_1 \geq B-\delta} e^{\mu x_1} k(|x|)dx\right\}$$

$$\leq \frac{1}{\mu} \log\left\{g'(0)e^{\mu(B-\delta)} + \tfrac{1}{2}e^{\mu B}\right\}$$

$$= B + \frac{1}{\mu} \log\left\{g'(0)e^{-\mu\delta} + \tfrac{1}{2}\right\} .$$

Thus if μ is so large that $g'(0)e^{-\mu\delta} < \frac{1}{2}$, we have

$$\Phi(\mu) < B . \tag{4.9}$$

If we let μ approach infinity in (4.8) and recall that δ is arbitrarily small, we find that

$$\lim_{\mu\to\infty} \Phi(\mu) = B$$

The inequality (4.9) now implies that for some values of μ, $\Phi(\mu)$ is smaller than its limits at 0 and $+\infty$. Since Φ is continuous, it attains its minimum value c^* at a unique point μ^*.

Since $\Phi'(\mu^*) = 0$, (4.4) follows from (4.5). Since $\mu^2\Phi'$ is increasing, it must be negative for $\mu < \mu^*$, which proves the last statement of the lemma.

We see from this proof that $c^* < B$. Moreover, since $\psi(\mu)$ is increasing and $\Psi(0) = 0$ by symmetry, c^* is always positive. Thus the asymptotic speed c^* is positive and smaller than the maximum speed of migration B.

In order to prove Theorem 1, we let $w_n(x)$ be the solution of the linear recursion

$$w_{n+1} = R[w_n] , \qquad (4.10)$$
$$w_0 = u_0 ,$$

where

$$R[w](x) \equiv g'(0) \int_{R^N} k(|y|)w(x-y)dy.$$

Then R has the monotonicity property (3.11), where S is the set of all bounded nonnegative functions on R^N. Moreover, we see from (3.5) that the recursion $u_{n+1} = Q[u_n]$ implies $u_{n+1} \le R[u_n]$. Thus Proposition 1 shows that

$$u_n(x) \le w_n(x) \qquad (4.11)$$

for all n, so that the function w_n gives a bound for u_n.

We now introduce the new functions

$$p_n(x) = e^{\mu^*(x_1 - nc^*)} w_n(x) . \qquad (4.12)$$

Substituting in (4.10), we find that these functions satisfy the recursion

$$p_{n+1}(x) = \int K(x-y)p_n(y)dy \tag{4.13}$$

$$p_0(x) = e^{\mu^* x_1} u_0(x) \quad ,$$

where we have defined the new kernel

$$K(x) = g'(0)e^{\mu^*(x_1 - c^*)} k(|x|) \quad . \tag{4.14}$$

Because $\Phi(\mu^*) = c^*$, the definition (4.1) of Φ shows that

$$\int K(x)dx = 1 \tag{4.15}$$

We can solve the convolution recursion (4.13) by means of Fourier transforms. Denoting the transforms of K and p_n by $\hat{K}$ and $\hat{p}_n$, respectively, we find that

$$\hat{p}_n = \hat{K}^n \hat{p}_0$$

It follows from (4.15) that

$$\hat{K}(0) = 1$$

$$|\hat{K}(\omega)| < 1 \qquad \text{for} \quad \omega \neq 0 \quad .$$

Because $k(|x|)$ is continuous and vanishes for $|x| \geq B$, $K(x)$ is square integrable. Hence by Parseval's equation $|\hat{K}(\omega)|^2$ has a finite integral. Since $|\hat{K}|^2$ for $n \geq 2$ and since

$$\lim_{n\to\infty} |\hat{K}(\omega)^n| = 0$$

for $\omega \neq 0$, we find from the dominated convergence theorem that

$$\epsilon_n \equiv (2\pi)^{-N} \int |\hat{K}|^n d\omega$$

approaches zero as $n \to \infty$. The Fourier inversion theorem shows that

$$p_n(x) \le \varepsilon_n \int p_0(y)\,dy$$
$$= \varepsilon_n \int u_0(y)\, e^{\mu^* y_1}\, dy \ .$$

Returning to u_n via (4.12) and (4.11), we see that

$$u_n(x) \le \varepsilon_n \int u_0(y) e^{\mu^* y_1}\, dy \qquad \text{for } x_1 \ge nc^* \ .$$

By rotating the coordinate axes, we can make any direction the x_1-direction. That is, we may choose any unit vector ν and replace x_1 by $x \cdot \nu$. Because of the rotational symmetry of $k(|x|)$, ε_n , the integral of $(2\pi)^{-N}|\hat{K}|^n$, does not depend on ν . Thus we find that

$$u_n(x) \le \varepsilon_n \int u_0(y) e^{\mu^* y \cdot \nu}\, dy \qquad \text{for } x \cdot \nu \ge nc^* \ .$$

For any x with $|x| \ge nc^*$ we choose $\nu = x/|x|$ to find that

$$u_n(x) \le \varepsilon_n \int u_0 e^{\mu^* y \cdot x/|x|}\, dy \qquad \text{for } |x| \ge nc^* \ .$$

Because ε_n converges to 0 as $n \to \infty$, this inequality implies (3.8) and therefore proves Theorem 1.

REMARKS. 1. We see from the proof that the condition that u_0 vanish outside a bounded set can be replaced by the condition that the integral of $u_0(x) e^{\mu^* x \cdot \nu}$ is uniformly bounded for all unit vectors ν . If the integral is only bounded for a fixed unit vector ν , we find that the maximum of u_n on the set $x \cdot \nu \ge nc^*$ approaches zero.

2. If the condition in Remark 1 holds not for μ^* but for some positive $\mu < \mu^*$, the result is valid when c^* is replaced by $\Phi(\mu)$ in (3.8).

3. The argument used to show that ε_n approaches zero is a weak form of a local central limit theorem. A stronger version (see, e.g. Feller [3, p. 289, Theo. 2]) shows that ε_n is of the order $n^{-\frac{1}{2}N}$ as $n \to \infty$.

5. PROOF OF THEOREMS 2 AND 3

In order to prove Theorem 2 we need to bound u_n from below. We shall construct a sequence of comparison functions σp_n which serves to show that $u_n(x)$ is uniformly positive on a ball whose radius grows like nc_2 with some $c_2 > c$. From this it will follow that u_n is arbitrarily near one on a ball of radius nc when n is large.

We recall that the inequality $g'(0) > 1$ is an immediate consequence of the inequalities (3.4) and (3.5), and that this inequality was needed to prove Lemma 4.1. We shall continue to use the inequality $g'(0) > 1$ in this section, but we shall make no further use of (3.5).

We observe that if $h < g'(0)$, then $g(u) \geq hu$ for sufficiently small positive values of u, and define the approximation

$$\varphi(\mu) = \frac{1}{\mu} \log \left\{ h \int e^{\mu x_1} k(|x|) dx \right\} \tag{5.1}$$

to $\Phi(\mu)$. We shall again use the function

$$\psi(\mu) = \frac{\int x_1 e^{\mu x_1} k(|x|)dx}{\int e^{\mu x_1} k(|x|)dx} \tag{5.2}$$

and the inequality (4.3):

$$\psi'(\mu) > 0 \tag{5.3}$$

We need the following lemma:

LEMMA 5.1. *For any* c_1 *such that*

$$0 < c_1 < c^* \tag{5.4}$$

there are constants h *and* μ_1 *such that*

$$1 < h < g'(0) \tag{5.5}$$

and

$$\varphi(\mu_1) > \psi(\mu_1) > c_1 \quad . \tag{5.6}$$

Moreover,

$$\varphi'(\mu) < 0 \quad \text{for} \quad 0 < \mu \le \mu_1 \quad . \tag{5.7}$$

PROOF. Lemma 4.1 states that $\psi(\mu)$ is increasing and $\Phi(\mu)$ is decreasing for $\mu \le \mu^*$ and that $\Phi(\mu^*) = \psi(\mu^*) = c^* > c_1$. By continuity,

$$\Phi(\mu_1) > \Psi(\mu_1) > c_1$$

if $\mu_1 < \mu^*$ is sufficiently close to μ^* . Since $\varphi(\mu_1)$ approaches $\Phi(u_1)$ as h goes to $g'(0)$, (5.6) holds whenever $h < g'(0)$ is sufficiently close to $g'(0)$.

Differentiation shows that

$$\varphi'(\mu) = -\frac{1}{\mu}[\varphi(\mu) - \Psi(\mu)] \quad .$$

Therefore it follows from (5.6) that $\varphi'(\mu_1) < 0$. A second differentiation shows that

$$(\mu^2\varphi')' = \mu\Psi' > 0 \quad .$$

Consequently $\mu^2\varphi'$ is increasing, and (5.7) follows.

From now on we shall keep fixed values of μ_1 and h corresponding to a given $c_1 < c^*$ such that (5.5), (5.6), and (5.7) are satisfied. We define σ_0 to be the smallest positive root of the equation

$$g(\sigma) = h\sigma \quad .$$

Then

$$g(u) \geq hu \qquad \text{for} \quad 0 \leq u \leq \sigma_0 \tag{5.8}$$

We are now ready to construct a sequence of comparison functions which will give lower bounds for the u_n . Our construction will be done in two stages. We first construct a one-dimensional function and then use this function to construct a rotationally symmetric function of bounded support.

We begin with the family of one-dimensional functions

$$q(x_1) = \begin{cases} e^{-\mu x_1} \sin \eta x_1 & \text{for} \quad 0 \leq x_1 \leq \frac{\pi}{\eta} \\ 0 & \end{cases} \tag{5.9}$$

which depend upon the parameters μ and η .

LEMMA 5.2. <u>For any</u> c_1 <u>which satisfies</u> $0 < c_1 < c^*$ <u>there exist positive constants</u> η <u>and</u> μ <u>such that if</u>

$$0 \le \sigma \le \sigma_0 \quad , \tag{5.10}$$

<u>the function</u> $\sigma q(x_1)$ <u>satisfies the inequality</u>

$$Q[\sigma q] \ge \sigma q(x_1 - c_1) \quad . \tag{5.11}$$

PROOF. It is obvious from the translation invariance of Q that the left-hand side of (5.11) is again a function of x_1 only. Since $Q[\sigma q] \ge 0$, we need to verify (5.11) only for

$$c_1 < x_1 < c_1 + \frac{\pi}{\eta} \quad . \tag{5.12}$$

We see from (5.8) that since $\sigma q < \sigma \le \sigma_0$,

$$Q[\sigma q] \ge \sigma h \int\limits_{0 \le y_1 \le \pi/\eta} k(|x-y|)e^{-\mu y_1} \sin \eta y_1 dy \quad . \tag{5.13}$$

We shall suppose that η is chosen so small that

$$\frac{\pi}{\eta} > c_1 + B \quad .$$

Then for $c_1 < x_1 < c_1 + \frac{\pi}{\eta}$ the inequality $|x - y| \le B$ implies that

$$-\frac{\pi}{\eta} < y_1 < \frac{2\pi}{\eta} \quad .$$

For y_1 in this interval but outside the interval $[0, \frac{\pi}{\eta}]$, $\sin \eta y_1$ is negative. Since $k(|x-y|) = 0$ for $|x-y| > B$, we conclude that when $c_1 < x_1 < c_1 + \frac{\pi}{\eta}$, the integral on the right of (5.13) is decreased if the restriction $0 \le y_1 \le \frac{\pi}{\eta}$ is removed. That is,

$$Q[\sigma q] \ge \sigma h \int_{R^N} k(|x-y|)e^{-\mu y_1} \sin \eta y_1 dy \tag{5.14}$$

$$= \sigma h \int k(|y|)e^{-\mu(x_1-y_1)} \sin \eta(x_1-y_1)dy$$

$$= h \int k(|y|)e^{-\mu(c_1-y_1)} \cos \eta(c_1-y_1)dy\ \sigma e^{-\mu(x_1-c_1)} \sin \eta(x_1-c_1)$$

$$+ h \int k(|y|)e^{-\mu(c_1-y_1)} \sin \eta(c_1-y_1)dy \sigma e^{-\mu(x_1-c_1)} \cos \eta(x_1-c_1)$$

$$\text{for } c_1 \le x_1 \le c_1 + \frac{\pi}{\eta} .$$

Thus the inequality (5.11) will follow from the inequality

$$h \int k(|y|)e^{-\mu(c_1-y_1)} \cos \eta(c_1-y_1)dy > 1 \tag{5.15}$$

together with the equation

$$h \int k(|y|)e^{-\mu(c_1-y_1)} \sin \eta(c_1-y_1)dy = 0 . \tag{5.16}$$

We rewrite (5.16) for $\eta \ne 0$ in the form

$$H(\mu,\eta) \equiv \frac{\int k(|y|)e^{\mu y_1} \sin \eta(c_1-y_1)dy}{\eta \int k(|y|)e^{\mu y_1} dy} = 0 . \tag{5.17}$$

Taking the limit as η goes to zero, we define

$$H(\mu,0) = c_1 - \Psi(\mu) .$$

With this definition, H is analytic in μ and η for all real μ

and η. Since $\psi(0) = 0$, we see that $H(0,0) > 0$ while, by (5.6), $H(\mu_1,0) < 0$. Therefore there is a positive $\mu_0 < \mu_1$ such that

$$H(\mu_0,0) = 0 \quad .$$

In addition,

$$\frac{\partial H}{\partial \mu}(\mu_0,0) = -\psi'(\mu_0) < 0$$

by (5.3). The implicit function theorem now states that there is an $\eta_0 > 0$ such that for any η with $|\eta| \le \eta_0$ the equation (5.17) has a solution $\mu = \mu(\eta)$, $\mu(\eta)$ is a smooth function of η, and $\mu(0) = \mu_0$.

For $\eta = 0$, $\mu = \mu_0$ the left-hand side of (5.15) is equal to

$$e^{\mu_0[\varphi(\mu_0)-c_1]} \quad ,$$

which is greater than 1 by (5.6) and (5.7). Hence the inequality (5.15) is satisfied for $\eta = 0$, $\mu = \mu_0$. By continuity, it is satisfied when $\mu = \mu(\eta)$, provided η is sufficiently small.

Thus for sufficiently small η and $\mu = \mu(\eta)$, the inequality (5.15) and the equation (5.16) are satisfied. If η is also so small that $\frac{\pi}{\eta} > c_1 + B$, then (5.14) holds, (5.11) follows, and Lemma (5.2) is proved.

We shall now use the function $q(x_1)$ to produce a sequence of rotationally symmetric comparison functions with bounded support. We suppose that we are given a positive number $c_2 < c^*$ and that

$$0 < c_2 < c_1 < c^* \quad . \tag{5.19}$$

We fix the constants h, σ_0, μ, and η as in Lemma 5.2, and choose a constant D which is so large that

$$D \geq B + \frac{B^2}{2(c_1 - c_2)} \quad . \tag{5.20}$$

We now define the sequence

$$s_n(x) = \max_{\tau \geq -D-nc_2} q(|x| + \tau) \tag{5.21}$$

where $q(x_1)$ is defined by (5.9). If we define

$$M = \max q(x_1)$$

and let ξ be the unique value of x_1 such that

$$q(\xi) = M ,$$

it is easily seen that

$$s_n(x) = \begin{cases} M & \text{for} \quad |x| \leq \xi + D + nc_2 \\ q(|x| - D - nc_2) & \text{for} \quad |x| \geq \xi + D + nc_2 \quad . \end{cases} \tag{5.22}$$

LEMMA 5.3. <u>If</u> D <u>satisfies</u> (5.20) <u>and</u> $\sigma \leq \sigma_0$, <u>then</u>

$$Q[\sigma s_{n+1}] \geq \sigma s_n \qquad \text{for} \quad n = 0,1,2, \ldots \tag{5.23}$$

PROOF. We first prove (5.23) for $n = 0$.

When $|x| \leq \xi + D - B$, the inequality $|x-y| \leq B$ implies that $|y| \leq \xi + D$ so that $s_0(y) = M$. Since $k(|x-y|) = 0$ for $|x-y| \geq B$ and the integral of k is 1, we see that

$$Q[\sigma s_0] = g(\sigma M) > \sigma M = \sigma s_1(x) \quad \text{for} \quad |x| \le \xi + D - B$$

It only remains to verify (5.23) for $|x| > \xi + D - B$. We note that if $|y| \le B$ and $|x| \ge \xi + D - B$, then

$$|x-y| = (|x|^2 - 2x \cdot y + |y|^2)^{\frac{1}{2}}$$

$$\le |x| \left\{1 + \frac{-2\,x \cdot y + |y|^2}{2|x|^2}\right\}$$

$$\le |x| - \frac{x \cdot y}{|x|} + \frac{B^2}{2(\xi + D - B)}$$

$$\le |x| - \frac{x \cdot y}{|x|} + c_1 - c_2$$

by (5.20). Since $s_0(x)$ and $k(x)$ depend only on $|x|$, the same is true of $Q[\sigma s_0]$. Therefore we may, without loss of generality, assume that x lies along the positive x_1 axis so that $x \cdot y/|x_1| = y_1$.

Since s_0 is a nonincreasing function of $|x|$, we find that for $|y| \le B$ and $|x| \ge \xi + D - B$,

$$s_0(|x-y|) \ge s_0(|x| - y_1 + c_1 - c_2)$$

$$\ge q(|x| - y_1 + c_1 - c_2 + \tau)$$

for all $\tau \ge -D$. The second inequality comes from definition (5.21) of s_0. Thus for $|x| \ge \xi + D - B$ and $\tau \ge -D$ we see from (5.11) that

$$Q[\sigma s_0] = \int k(|y|)g(\sigma s_0(|x-y|))\,dy$$

$$\ge \int k(|y|)g(\sigma q(|x| + c_1 - c_2 + \tau - y_1))\,dy$$

$$= Q[\sigma q](|x| + c_1 - c_2 + \tau)$$

$$\geq \sigma q(|x| - c_2 + \tau) \quad .$$

Maximizing the right-hand side among all $\tau \geq -D$, or, equivalently, among constants $\tau - c_2 \geq -D - c_2$, we see from (5.21) that

$$Q[\sigma s_0] \geq \sigma s_1(x) \quad .$$

Thus (5.23) holds for $n = 0$.

Since s_n is obtained from s_0 by replacing D by the larger constant $D + nc_2$, which also satisfies (5.20), the same proof gives (5.23) for all positive n, and the lemma is proved.

In order to use the functions σs_n as comparison functions, we shall show that for some integer ℓ $u_\ell(x)$ lies above a translate of σs_0. Since $s_0 = 0$ for $|x| \geq D + \frac{\pi}{\eta}$ and σ is arbitrarily small, it suffices to show that u_ℓ is uniformly positive on a ball of radius $D + \frac{\pi}{\eta}$.

Because $k(|x|)$ is continuous and has integral 1, it must be uniformly positive on some ball. That is, there are a point $\bar{x}$ and positive constants $\varkappa$ and R such that

$$k(|x|) \geq \varkappa > 0$$

for $|x - \bar{x}| \leq R$. Then if $|x - 2\bar{x}| < 3/2\ R$,

$$k^{(2)}(|x|) = \int k(|y|)k(|x-y|)\,dy \geq \varkappa^2 \int\limits_{\substack{|y-\bar{x}|<R \\ |x-y-\bar{x}|<R}} dy \geq \varkappa^2 \int\limits_{|y-\frac{1}{2}x| \leq \frac{1}{4}R} dy$$

$$= \varkappa^2 V_N\left(\tfrac{1}{4}R\right)^N$$

where V_N is the volume of the unit N-ball. If we define

$$k^{(\ell)}(|x|) = \int k(|y|)k^{(\ell-1)}(|x-y|)dy$$

recursively, we find by induction that

$$k^{(\ell)}(|x|) \geq \varkappa^{\ell}\left[V_N(\tfrac{1}{4}R)^N\right]^{\ell-1} \quad \text{for } |x-\ell\bar{x}| \leq \tfrac{1}{2}(\ell+1)R .$$

Since u_0 has a positive integral, there is a ball $|x-\hat{x}| < \frac{1}{2}R$ over which the integral of u_0 is positive. Because $g(u) \geq u$, we find that

$$u_\ell(x) \geq \int_{|y-\hat{x}|\leq R} k^{(\ell)}(|x-y|)u_0(y)dy$$

$$\geq \varkappa^{\ell}\left[V_N\left(\tfrac{1}{4}R\right)^N\right]^{\ell-1} \int_{|y-\hat{x}|\leq\frac{1}{2}R} u_0(y)dy$$

$$\text{for } |x - \ell\bar{x} - \hat{x}| \leq \tfrac{1}{2}\ell R .$$

That is, u_ℓ is bounded below on a ball of radius $\frac{1}{2}\ell R$. Given any positive $c < c^*$, we choose c_1 and c_2 so that

$$0 < c < c_2 < c_1 < c^* .$$

We then construct the functions of Lemma 5.3. We observe that $s_0 = 0$ for $|x| \geq D + \frac{\pi}{\eta}$ and choose ℓ so that

$$\tfrac{1}{2}\ell R \geq D + \frac{\pi}{\eta} .$$

Since $s_0 \leq M$, we must only choose

$$\sigma = \min\left\{\sigma_0,\ M^{-1}\varkappa^{\ell}\left[V_N\left(\tfrac{1}{4}R\right)^N\right]^{\ell-1} \int_{|z-\hat{x}|\leq\frac{1}{2}R} u_0(z)dz\right\}$$

to make

$$u_\ell(x + \ell\overline{x} + \hat{x}) \geq \sigma s_0(x) \quad .$$

Then by Proposition 1

$$u_{j+\ell}(x + \ell\overline{x} + \hat{x}) \geq \sigma s_j(x) \qquad \text{for } j = 1,2,\ldots \quad .$$

In particular, if j is so large that $D + jc_2 > |\ell\overline{x} + \hat{x}|$,

$$u_{j+\ell}(x) \geq \sigma M \quad \text{for} \quad |x| < D - |\ell\overline{x} + \hat{x}| + jc_2 \quad . \tag{5.24}$$

To prove Theorem 2 we need to show that for any positive ε and all sufficiently large n , $u_n(x) > 1 - \varepsilon$ for $|x| \geq nc$. To establish this fact we prove the following lemma.

LEMMA 5.3. <u>To any</u> $\varepsilon > 0$ <u>there corresponds an integer</u> m_ε <u>which is independent of</u> j <u>and</u> ℓ <u>such that if</u> (5.24) <u>is valid and if</u> $D - |\ell\overline{x} + \hat{x}| + jc_2 \geq Bm_\varepsilon$, <u>then</u>

$$u_{j+\ell+m_\varepsilon} \geq 1 - \varepsilon \qquad \text{for} \quad |x| \leq D - |\ell\overline{x} + \hat{x}| - Bm_\varepsilon + jc_2 \quad . \tag{5.25}$$

PROOF. We define a sequence of constants λ_n by

$$\lambda_{n+1} = g(\lambda_n) \quad , \tag{5.26}$$

$$\lambda_0 = \sigma M \quad .$$

Since $g(u) > u$ for u in $(0,1)$, λ_n increases to 1 as $n \to \infty$. Hence for any $\varepsilon > 0$ there is an m_ε such that

$$\lambda_{m_\varepsilon} > 1 - \varepsilon \quad .$$

Suppose that for some nonnegative integer m

$$u_{j+\ell+m} \geq \lambda_m \qquad \text{for} \quad |x| \leq D - |\ell\bar{x}+\hat{x}| + jc_2 - mB \;. \quad (5.27)$$

Since $k(|x-y|) = 0$ for $|x-y| \geq B$ and since the inequalities $|x-y| \leq B$ and $|x| \leq D - |\ell\bar{x}+\hat{x}| + jc_2 - (m+1)B$ imply that $|y| \leq D - |\ell\bar{x}+\hat{x}| + jc_2 - mB$, we find that

$$u_{j+\ell+m+1} = Q\left[u_{j+\ell+m}\right] \geq g(\lambda_m) = \lambda_{m+1}$$

$$\text{for } |x| \leq D - |\ell\bar{x}+\hat{x}| + jc_2 - (m+1)B \quad .$$

Since (5.27) reduces to (5.24) for $m = 0$, we have proved (5.27) by induction. By putting $m = m_\varepsilon$ so that $\lambda_m \geq 1-\varepsilon$, we obtain the statement (5.25) of the Lemma.

To finish the proof of Theorem 2 we set $n = j+\ell+m_\varepsilon$ in (5.25) to see that for all sufficiently large n

$$u_n(x) \geq 1-\varepsilon \quad \text{for} \quad |x| \leq D - |\ell\bar{x}+\hat{x}| - m_\varepsilon B + (n-\ell-m_\varepsilon)c_2 \quad .$$

Since $c < c_2$, we can find an n_ε such that

$$D - |\ell\bar{x}+\hat{x}| - m_\varepsilon B + (n-\ell-m_\varepsilon)c_2 \geq nc \qquad \text{for} \quad n \geq n_\varepsilon \quad .$$

Then

$$u_n(x) \geq 1-\varepsilon \qquad \text{for} \quad n \geq n_\varepsilon \quad , \quad |x| \leq nc \quad .$$

Since ε is arbitrary, this proves Theorem 2.

To prove Theorem 3 we choose any c_1 and c_2 such that $0 \leq c_2 < c_1 < c^*$ and construct the functions s_n in Lemma 5.3.

By hypothesis, $u_0 \geq \rho_0$ for $|x| \geq b$. Therefore, if we choose σ so that $\sigma \leq \sigma_0$ and

$$\sigma M \leq \rho_0 \quad ,$$

we find that

$$u_0(x+z) \geq \sigma\, s_0(x) \quad \text{whenever} \quad |z| > b + D + \frac{\pi}{\eta} \quad .$$

We choose ℓ so large that

$$\ell c_2 \geq b + \frac{\pi}{\eta} - \xi \quad . \tag{5.28}$$

By Lemma 5.3 and Proposition 1 we find that

$$u_\ell(y) \geq \sigma\, s_\ell(|y-z|)$$

$$= \sigma M \quad \text{for} \quad |y-z| \leq \xi + D + \ell c_2 \; , \; |z| \geq b + D + \frac{\pi}{\eta} \; .$$

For $|y| \geq b + D + \frac{\pi}{\eta}$ we choose $z = y$ to see that the inequalities on $|y-z|$ and $|z|$ are satisfied, so that

$$u_\ell(y) \geq \sigma M \quad . \tag{5.29}$$

For $0 < |y| < B + D + \frac{\pi}{\eta}$, we let

$$z = \frac{1}{|y|}\left(b + D + \frac{\pi}{\eta}\right) y$$

to see that by (5.28)

$$|y - z| = b + D + \frac{\pi}{\eta} - |y| \leq \xi + D + \ell c_2 \quad .$$

Since $|z| = b + D + \frac{\pi}{\eta}$, (5.29) still holds. For $y = 0$ we reach the same conclusion by choosing any z with $|z| = b + D + \frac{\pi}{\eta}$.

Thus $u_\ell(y)$ is uniformly bounded below by the positive constant σM .

Proposition 1 now shows that

$$u_{\ell+n} \geq \lambda_n$$

where the sequence of constants λ_n is defined by (5.26). Since λ_n increases to 1 as $n \to \infty$, this inequality proves Theorem 3.

6. LARGE-TIME BEHAVIOR IN THE HETEROZYGOTE SUPERIOR CASE

In the heterozygote superior case (2.9 ii) the function $g(u)$ is equal to u at

$$u^* = \frac{\beta - \gamma}{2\beta - \alpha - \gamma} .$$

It still has the properties (3.3) and (3.5). In place of (3.4), we now have

$$g(u) > u \quad \text{for} \quad 0 < u < u^* \tag{6.1}$$

and

$$g(u) < u \quad \text{for} \quad u^* < u < 1 . \tag{6.2}$$

Moreover,

$$g'(0) > 1 , \quad g'(1) < 1 . \tag{6.3}$$

Since the proof of Theorem 1 used only the hypotheses (3.3) and (3.5), Theorem 1 is valid for this case.

The inequalities (6.1) and (6.3) state that the function $g(u)$ satisfies the hypotheses of Theorem 2 for u on the interval $[0,u^*]$.

Therefore the proof of Theorem 2 shows that if u_0 is not identically zero and if $0 < c < c^*$, then

$$\liminf_{n\to\infty} \min_{|x|\le nc} u_n(x) \ge u^* .$$

On the other hand, (6.2) and (6.3) show that the function $1 - g(1-v)$, which occurs in the recursion for the new variable $v_n \equiv 1 - u_n$, satisfies the hypotheses of Theorem 3 for $0 \le v \le 1 - u^*$. Therefore the proof of Theorem 3 shows that if u_0 has bounded support, then

$$\limsup_{n\to\infty} \max_{x} u_n(x) \le u^* .$$

By putting these results together, we obtain the following theorem.

THEOREM 4. *Let* $g(u)$ *satisfy the conditions* (3.3), (3.5), (6.1), (6.2), *and* (6.3) *of the heterozygote superior case. If* $u_0(x)$ *has bounded support and if its integral is not zero, then*

$$\lim_{n\to\infty} \max_{|x|\ge nc^*} u_n(x) = 0 ,$$

$$\lim_{n\to\infty} \max_{R^N} u_n(x) = u^* ,$$

and for every $c < c^*$

$$\lim_{n\to\infty} \min_{|x|\le nc} u_n(x) = u^* .$$

7. THE EXISTENCE OF TRAVELLING WAVES

A nonconstant solution of the recursion $u_{n+1} = Q[u_n]$ which is of the form

$$u_n(x) = W(x \cdot \nu - nc) \quad ,$$

where ν is a fixed unit vector, is called a travelling wave with speed c. For each n the travelling wave depends only on the single variable $x \cdot \nu$, and the solution is translated a distance c in the direction of ν at each time step. Substituting $W(x \cdot \nu - nc)$ into the recursion, we see that it is a travelling wave if and only if

$$Q[W(x \cdot \nu)] = W(x \cdot \nu - c)$$

or equivalently

$$Q[W](x + c\nu) = W \quad .$$

That is, the function W of one variable gives a travelling wave if and only if it is a fixed point of the operator $Q[u](x + c\nu)$.

Because Q is rotationally invariant, we see that if, for some unit vector ν, $W(x \cdot \nu - nc)$ is a travelling wave solution, then so is $W(x \cdot \tau - nc)$ for any other unit vector τ. For this reason we shall, without loss of generality, choose ν to lie along the positive x_1-axis, so that $x \cdot \nu = x_1$. Thus our travelling wave will be of the form $W(x_1 - nc)$, and $W(x_1 - nc)$ is a travelling wave of speed c if and only if W is a fixed point of the operator

$$Q_c[u] \equiv Q[u](x_1 + c, x_2, \ldots, x_N) \quad . \tag{7.1}$$

We still assume that Q is of the form

$$Q[u] = \int_{R^N} k(|x-y|)\, g(u(y))\,dy$$

where $k(|y|)$ is a continuous nonnegative function whose integral over R^N is 1 and which vanishes for $|x| \geq B$. We continue to assume that g is differentiable and nondecreasing, that $g(0) = 0$ and $g(1) = 1$, and that

$$g(u) \leq g'(0)u \qquad \text{for} \quad 0 \leq u \leq 1 \quad .$$

In addition we now suppose that there is a constant D such that

$$g'(0)\left[u - Du^2\right] \leq g(u) \leq g'(0)u \quad . \tag{7.2}$$

When g is of the form (2.4), these hypotheses are satisfied in the heterozygote superior case (2.9 ii), and in the special part of the heterozygote intermediate case where the inequality (2.10) holds, which was treated in Theorems 1 to 3.

It is clear from the translation invariance of Q that if $w(x_1)$ depends only on x_1, the same is true of $Q_c[w]$.

As before, we define the asymptotic speed c^* by (3.6). For $c > c^*$ we denote by $\mu(c)$ the root below μ^* of the equation $\Phi(\mu) = c$. Our existence theorem is the following.

THEOREM 5. <u>For each $c \geq c^*$ there exists a travelling wave solution $W(x_1 - nc)$ of speed c such that $W(x_1)$ is nonincreasing and</u>

$$W(-\infty) = 1 , \quad W(\infty) = 0 . \tag{7.3}$$

<u>For</u> $c > c^*$

$$\lim_{x_1 \to +\infty} e^{\mu(c)x_1} W(x_1) = 1 , \tag{7.4}$$

<u>while</u> <u>for</u> $c = c^*$

$$\lim_{x_1 \to +\infty} \frac{1}{x_1} e^{\mu^* x_1} W(x_1) = 1 . \tag{7.5}$$

PROOF. We first suppose that $c > c^*$. We define the function

$$z_0(x_1) = \begin{cases} 1 & \text{for } x_1 < 0 \\ e^{-\mu(c)x_1} & \text{for } x_1 \geq 0 . \end{cases}$$

Then

$$\begin{aligned} Q[z_0] &\leq g'(0) \int k(|y|) z_0(x-y)\,dy \\ &\leq g'(0) \int k(|y|) e^{-\mu(c)(x_1-y_1)} dy \\ &= e^{-\mu(c)[x_1 - \Phi(\mu(c))]} \\ &= e^{-\mu(c)(x_1-c)} . \end{aligned}$$

Since also $Q[z_0] \leq 1$, we find that

$$Q[z_0] \leq z_0(x_1 - c) .$$

Thus

$$Q_c[z_0] \leq z_0 . \tag{7.6}$$

We now define the recursion

$$z_{n+1} = Q_c[z_n] \tag{7.7}$$

with z_0 as above. By applying Proposition 1 with $v_n = z_{n+1}$ and $w_n = z_n$, we find that

$$z_{n+1} \leq z_n$$

for all n. Thus $z_n(x)$ is a nonincreasing sequence of nonnegative functions. Consequently, it has a limit $W(x_1)$ as $n \to \infty$. The dominated convergence theorem shows that the right-hand side of (7.7) converges to $Q_c[W]$. Hence W is a solution of

$$W = Q_c[W] .$$

Since z_0 is a nonincreasing function of x_1, we have $z_0(x_1) \geq z_0(x_1 + b)$ for any positive b. Since the operator Q_c is translation invariant, it follows from Proposition 1 that for any n, $z_n(x_1) \geq z_n(x_1 + b)$ when $b > 0$. That is, each z_n is nonincreasing. Therefore, the same is true of the limit function $W(x_1)$.

We need to show that $W(x_1)$ is not constant. Because $W \leq z_0$ and z_0 approaches zero as $x_1 \to \infty$, $W(+\infty) = 0$, and the only possible constant would be 0.

In order to prove that W is not identically zero, we choose a constant λ such that

$$\mu(c) < \lambda \leq \min(\mu^*, 2\mu(c))$$

and a positive constant τ, and define the function

$$v(x_1) = \begin{cases} 0 & \text{for } x_1 \le 0 \\ \tau\left(e^{-\mu(c)x_1} - e^{-\lambda x_1}\right) & \text{for } x_1 \ge 0 \end{cases} .$$

Then

$$g'(0) \int k(|y|)v(x_1-y_1)dy \ \ge\ \tau g'(0) \int k(|y|)\left[e^{-\mu(c)(x_1-y_1)} - e^{-\lambda(x_1-y_1)}\right]dy$$

$$= \tau\left(e^{-\mu(c)(x_1-c)} - e^{-\lambda(x_1-\Phi(\lambda))}\right)$$

$$= v(x_1 - c) + \tau e^{-\lambda(x_1-c)}\left\{1 - e^{-\lambda[c-\Phi(\lambda)]}\right\} .$$

We therefore see from the hypothesis (7.2) that for any τ such that

$$0 < \tau \le 1$$

we have for $x_1 \ge c$

$$Q[v] \ge v(x_1-c) + \tau e^{-\lambda(x_1-c)}\left\{1 - e^{-\lambda[c-\Phi(\lambda)]} - \tau D e^{-[2\mu-\lambda]x_1+2\mu\Phi(2\mu)-\lambda c}\right\}$$

$$\ge v(x_1-c) + \tau e^{-\lambda(x_1-c)}\left\{1 - e^{-\lambda[c-\Phi(\lambda)]} - \tau D e^{-2\mu[c-\Phi(2\mu)]}\right\} .$$

We have written μ for $\mu(c)$, and used the facts that $\lambda \le 2\mu$ and $x_1 \ge c$ in obtaining the second inequality.

By Lemma 4.1, $\Phi(\mu)$ is decreasing on the interval $(0, \mu^*)$. Since $\Phi(\mu(c)) = c$ and $\mu(c) < \lambda \le \mu^*$, we see that $\Phi(\lambda) < c$, so that

$$e^{-\lambda[c-\Phi(\lambda)]} < 1 \quad .$$

Therefore if we make τ sufficiently small, the second term on the right of (7.8) is nonnegative and hence

$$Q[v] \geq v(x_1 - c)$$

for $x_1 \geq c$. Since $v(x_1 - c) = 0$ for $x_1 < c$, this inequality holds for all x_1 . We write it in the equivalent form

$$Q_c[v] \geq v \quad . \tag{7.9}$$

We now define

$$v_n(x) = v\left(x_1 - \frac{1}{\mu(c)} \log \frac{1}{\tau}\right)$$

for all n . We see from the definitions of $z_0(x)$ and $v(x)$ that

$$z_0 \geq v_0 \quad .$$

By definition, $z_{n+1} = Q_c[z_n]$, while it follows from (7.9) that $v_{n+1} \leq Q_c[v_n]$. Thus Proposition 1 states that $z_n \geq v_n$ for all n . Letting n approach infinity, we find that

$$W(x_1) \geq v\left(x_1 - \frac{1}{\mu(c)} \log \frac{1}{\tau}\right) \quad .$$

This clearly shows that W is not identically zero. The asymptotic property (7.4) is a direct consequence of the fact that

$$v(x_1) \leq W(x_1) \leq z_0(x_1) \quad .$$

The fact that $W(-\infty) = 1$ follows by applying Theorem 2 to the travelling wave solution $W(x_1-nc)$. Thus Theorem 5 is established for $c > c^*$.

When $c = c^*$, we replace the function z_0 by

$$z_0 = \begin{cases} \left(x_1 - E + e^{\mu^* E}\right) e^{-\mu^* x_1} & \text{for } x_1 \geq E \\ 1 & \text{for } x_1 \leq E \end{cases}$$

where E is a constant which satisfies the inequalities

$$e^{\mu^* E} \geq \max\left\{\frac{1}{\mu^*}, 1 + E\right\} \tag{7.10}$$

and

$$E \geq B \quad . \tag{7.11}$$

It follows from (7.10) that $0 \leq z_0 \leq 1$. Moreover, it is easily verified that $\left(x_1 - E + e^{\mu^* E}\right) e^{-\mu^* x_1} > 1$ for $0 < x_1 < E$. Since $k(|x-y|) = 0$ for $|x-y| > B$, we then see from (7.11) that for $x_1 \geq E$

$$\begin{aligned} Q[z_0] &\leq g'(0) \int k(|x-y|)\left(y_1 - E + e^{\mu^* E}\right) e^{-\mu^* y_1}\, dy \\ &= g'(0) \int k(|y|)\left(x_1 - y_1 - E + e^{\mu^* E}\right) e^{-\mu^*(x_1 - y_1)}\, dy \\ &= \left(x_1 - \Psi(\mu^*) - E + e^{\mu^* E}\right) e^{-\mu^*(x_1 - \Phi(\mu^*))} \\ &= z_0(x_1 - c^*) \end{aligned}$$

by (4.4). Since $Q_{c^*}[z_0] \leq 1$, we see that

$$Q_{c^*}[z_0] \leq z_0$$

for all x. Thus, we have (7.6) with $c = c^*$, and we construct the nondecreasing solution W of $Q_{c^*}[W] = W$ as the limit of the recursion (7.7) as before.

To obtain a lower bound for W, we replace the function v by

$$v(x_1) = \begin{cases} \tau\left\{x_1 e^{-\mu^* x_1} - \dfrac{B(e^{-\mu^* x_1} - e^{-3\mu^* x_1/2})}{e^{\mu^* B/2} - 1}\right\} & \text{for } x_1 > 0 \\ 0 & \text{for } x_1 < 0 \end{cases}$$

It is easily seen that the function in the first line of this formula is positive for $x_1 > 0$ and negative for $-B < x_1 < 0$. An argument which uses this fact together with (4.4) and (7.2) shows that $Q_{c^*}[v] \geq v$ when τ is sufficiently small. As before, the inequality

$$v(x_1 - \frac{1}{\mu^*} \log \frac{1}{\tau}) \leq W(x_1) \leq z_0(x_1)$$

follows. This proves that W is not zero and gives the asymptotic formula (7.5). Theorem 2 again shows that $W(-\infty) = 1$, and the theorem is proved.

It is, of course, an immediate consequence of Theorem 2 that there can be no travelling wave with speed c when $|c| < c^*$.

It would be interesting to know whether, for a fixed $c \geq c^*$, every travelling wave of speed c is a translate of the function $W(x_1 - nc)$ which we have constructed. The existence of the travelling wave for $c > c^*$ has been proved independently by O. Diekmann [Thresholds and travelling waves for the geographical spread of infection. Preprint, Mathematisch Centrum, Amsterdam, 1977].

8. RELATIONS WITH THE FISHER EQUATION

We wish to examine some connections between the model presented in Section 2 and the Fisher model.

If the time of one life cycle in the model of Section 2 is denoted by τ , it is reasonable to write

$$u(x, n\tau) \equiv u_n(x)$$

and to define $u(x,t)$ for all t by linear interpolation for those times t which are not multiples of τ .

If τ is small, we can expect that the survival rates α , β , and γ will be close to 1 , and that the migration will not get very far. We consider a one-parameter family of recursion models parametrized by τ in which

$$\alpha = 1 - \hat{\alpha}\tau + o(\tau) \quad ,$$

$$\beta = 1 - \hat{\beta}\tau + o(\tau) \quad ,$$

$$\gamma = 1 - \hat{\gamma}\tau + o(\tau) \quad ,$$

and $k(|x|) = \tau^{-N/2}\ell(\tau^{-\frac{1}{2}}|x|)$, where $\hat{\alpha}$, $\hat{\beta}$, and $\hat{\gamma}$ are death rates and $\ell(|x|)$ is a fixed probability kernel. Then

$$g(u) = u + \tau f(u) + o(\tau)$$

where

$$f(u) = u(1-u)[\hat{\gamma} - \hat{\beta} + (2\hat{\beta} - \hat{\alpha} - \hat{\gamma})u] \quad , \qquad (8.1)$$

and for any smooth function $\varphi(x)$

$$\int k(|x-y|)\varphi(y)\,dy = \varphi(x) + \frac{\tau}{2N}\Delta\varphi(x)\int \ell(|z|)|z|^2 dz + o(\tau) \quad .$$

Here Δ is the usual Laplace operator. Thus, if we set

$$D = \frac{1}{2N}\int \ell(|z|)|z|^2 dz$$

we find that

$$Q[\varphi] = \varphi + \tau[f(\varphi) + D\Delta\varphi] + o(\tau) \quad .$$

The recursion $u_{n+1} = Q[u_n]$ becomes

$$\frac{u(x,(n+1)\tau) - u(x,\tau)}{\tau} = D\Delta u + f(u) \quad .$$

By letting τ approach zero, we formally obtain the equation

$$\frac{\partial u}{\partial t} = D\Delta u + f(u) \quad , \tag{8.2}$$

which is Fisher's equation with the diffusion constant D .

It must, however, be remembered that our model assumes that the entire population reproduces and dies in each time interval of length τ . Thus the limit as τ approaches zero is a better model for a slapstick comedy than for a biological system.

While the above limit can be mathematically justified on a finite time interval, it is not clear that one can take limits of large-time asymptotic behavior. Nevertheless, the following argument shows that one can expect to obtain the asymptotic properties of the solutions of the Fisher equation from those of a recursion of the form $u_{n+1} = Q[u_n]$.

Define the operator $\hat{Q}$ as

$$\hat{Q}[\varphi] = v(x,1) \ ,$$

where $v(x,t)$ is the solution of the initial value problem

$$\frac{\partial v}{\partial t} = \Delta v + f(v) \qquad \text{for} \quad t > 0$$

$$v(x,0) = \varphi(x) \ .$$

If we define

$$u_n(x) = v(x,n) \ ,$$

then clearly

$$u_{n+1} = \hat{Q}[u_n] \ .$$

The operator $\hat{Q}$ is not of the form which we have discussed above. However, if we linearize $\hat{Q}$ about $u = 0$, we find that

$$\hat{Q}[u] = L[u] + O(\|u\|^2) \ ,$$

where $L|\varphi|$ is the value at $t = 1$ of the solution w of the linearized equation

$$\frac{\partial w}{\partial t} = \Delta w + f'(0)w \ ,$$

$$w(x,0) = \varphi(x) \ .$$

Thus,

$$L[\varphi] = \int \ell(|x-y|)\, \varphi(y)\,dy$$

where

$$\ell(|x|) = (4\pi)^{-N/2} e^{f'(0) - \frac{|x|^2}{4}} \quad . \qquad (8.3)$$

We assume that $f(0) = f(1) = 0$, so that $\hat{Q}[0] = 0$, $\hat{Q}[1] = 1$. If

$$f(u) \leq f'(0)u \quad \text{for} \quad 0 \leq u \leq 1 \ ,$$

then

$$\hat{Q}[\varphi] \leq L[\varphi]$$

for all φ with values in $[0,1]$. If, moreover, $f(u) > 0$ in $(0,1)$, we can extend Theorems 1, 2, 3, and 5 to the operator $\hat{Q}$, while if $f(u) > 0$ in $(0,u^*)$, $f(u) < 0$ in $(u^*,1)$, and $f'(1) < 0$, we can prove Theorems 4 and 5. If $f(u)$ is of the form (8.1), then the first condition is satisfied when

$$\hat{\alpha} \leq \hat{\beta} \leq \frac{1}{3}(\hat{\alpha} + 2\hat{\gamma}) \quad , \qquad a < \gamma \ , \qquad (8.4)$$

while the second condition is true when

$$\hat{\beta} < \hat{\alpha} \leq \hat{\gamma} \quad .$$

The latter is just the heterozygote superior case, while (8.4) is the linearized version of (2.10).

In proving these results it is only necessary to replace the linearization

$$R[u] = \int k(|x-y|)g'(0)u(y)\,dy$$

of $Q[u]$ about $u = 0$ by the linearization L of $\hat{Q}$. That is, one replaces the kernel $g'(0)k(|y|)$ by the kernel $\ell(|y|)$ in

(8.3). (One must, of course, extend the arguments to take account of the fact that $\ell(|y|)$ does not have bounded support, but that it is rapidly decaying.)

If this replacement is made in (3.6), one finds the asymptotic speed

$$c^* = \min_{\mu>0} \frac{1}{\mu} \log \left\{ \int \ell(|y|)\, e^{\mu y_1}\, dy \right\}$$

$$= \min_{\mu>0} \frac{1}{\mu} \left[f'(0) + \mu^2 \right]$$

$$= 2 \sqrt{f'(0)} \quad .$$

This propagation speed for the Fisher equation with $D = 1$ was already found by Fisher [6], by Kolmogoroff, Petrovsky, and Piscounoff [11], and by Kanel' [9, 10].

The results about Fisher's equation which are obtained in this way are, of course, already known [1, 2]. However, our methods will produce the same results for the case of Fisher's equation (8.2) in which the diffusion constant D and the growth function f have explicit periodic dependence on the time t, so that seasonal variations in growth, migration, and death rates can be taken into account. In such a case the function $u_n(x)$ represents the gene fraction at a particular but arbitrary instant of the n th growth cycle. A travelling wave solution for such a model represents periodic motion composed with a uniform motion in the x_1-direction.

BIBLIOGRAPHY

1. ARONSON, D.G. and WEINBERGER, H.F., Nonlinear diffusion in population genetics, combustion, and nerve pulse propagation, Partial Differential Equations and Related Topics, Lecture Notes in Mathematics, vol.446, Springer, 1975, pp. 5-49.

2. ARONSON, D.G. and WEINBERGER, H.F., Multidimensional non-linear diffusion arising in population genetics, Advances in Math. (in print).

3. FELLER, W., An Introduction to Probability Theory and its Applications, vol. II, Wiley, 1966.

4. FIFE, P.C., and McLEOD, J.B., The approach of solutions of non-linear diffusion equations to travelling wave solutions, Bull. Amer. Math. Soc. 81 (1975), pp. 1076-1078.

5. FIFE, P.C. and PELETIER, L.A., Nonlinear diffusion in population genetics, Arch. for Rat. Mech. and Anal. 64 (1977), pp. 93-110.

6. FISHER, R.A., The advance of advantageous genes, Ann. of Eugenics 7 (1937), pp. 355-369.

7. FLEMING, W.H., A selection-migration model in population genetics, J. Math. Biol. 2(1975), pp. 219-233.

8. HAMMERSLEY, J.M., Postulates for subadditive processes, Annals of Probability 2 (1974), pp. 652-680.

9. KANEL', JA.I., Stabilization of solutions of the Cauchy problem encountered in combustion theory, Mat. Sbornik (N.S.) 59 (101) (1962), supplement, pp. 245-288.

10. KANEL', JA.I., On the stability of solutions of the equation of combustion theory for finite initial functions, Mat. Sbornik (N.S.) 65 (107) (1964), pp. 398-413.

11. KOLMOGOROFF, A., PETROVSKY, I., and PISCOUNOFF, N., Étude de l'équation de la diffusion avec croissance de la quantité de matière et son application à un problème biologique, Bull. Univ. Moskou, Ser. Internat., Sec. A, 1 (1937) #6, pp. 1-37.

12. NAGYLAKI, T., Conditions for the existence of clines, Genetics 80 (1975), pp. 595-615.

13. SLATKIN, M., Gene flow and selection in a cline, Genetics 75 (1973), pp. 733-756.

A MINIMAX PRINCIPLE AND APPLICATIONS TO ELLIPTIC PARTIAL DIFFERENTIAL EQUATIONS

Paul H. Rabinowitz*

Mathematics Department
University of Wisconsin
Madison, Wisconsin 53706

INTRODUCTION

Let E be a real Banach space and I a continuously differentiable map from E to $\mathbb{R}$, i.e. $I \in C^1(E, \mathbb{R})$. The purpose of these lectures is to describe a minimax principle which can be used to determine critical points of I. Applications of this principle will be made to several abstract situations as well as to more concrete problems involving semilinear elliptic partial differential equations.

The basic ideas for the minimax principle go back to early work of Ljusternick and Schnirelman as well as to Morse [1]. As an interesting application, Ljusternick and Schnirelman showed in particular that if

* This research was sponsored in part by the Office of Naval Research under Contract No. N00014-76-C-0300 and by the U.S. Army under Contract No. DAAG-29-75-C-0024. Any reproduction in part or in full for the purposes of the U.S. Government is permitted.

$g \in C^1(\mathbb{R}^n, \mathbb{R})$ and g is even, i.e. $g(x) = g(-x)$, then $g|_{S^{n-1}}$ possesses at least n distinct pairs of critical points. This is surprising since without the evenness condition, you would only expect a maximum and minimum for $g|_{S^{n-1}}$.

In §1, we present the minimax principle together with some applications where one (nontrivial) critical point of I is obtained. In §2, some more subtle applications will be carried out. These involve multiple critical values of I such as the Ljusternick and Schnirelman theorem mentioned above.

§1. THE MINIMAX PRINCIPLE

Below E always denotes a real Banach space and $I \in C^1(E, \mathbb{R})$. We say I satisfies the Palais-Smale condition (P-S) if each sequence (u_m) such that $I(u_m)$ is bounded and $I'(u_m) \to 0$ is precompact. Here $I'(u)$ denotes the Fréchet derivative of I at $u \in E$. Note that this is a linear map from E to $\mathbb{R}$, i.e. $I'(u) \in E^*$, the dual space of E. The (PS) condition can be interpreted as a compactness condition. It implies in particular that for any $a < b$, $\{u \in E \mid a \le I(u) \le b \text{ and } I'(u) = 0\}$ is compact.

To present the minimax principle, we need a preliminary result. Let $c \in \mathbb{R}$, $K_c = \{u \in E \mid I(u) = c \text{ and } I'(u) = 0\}$, and $A_c = \{u \in E \mid I(u) \le c\}$.

LEMMA 1.1. <u>Suppose</u> $I \in C^1(E, \mathbb{R})$ <u>and satisfy</u> (P-S). <u>Let</u> $c \in \mathbb{R}$ <u>and</u> $\mathcal{O}$ <u>be any neighborhood of</u> K_c. <u>Then for any</u> $\bar{\varepsilon} > 0$, <u>there exists</u> $\varepsilon \in (0, \bar{\varepsilon})$ <u>and</u> $\eta \in C([0,1] \times E, E)$ <u>such that</u>:

(1°) $\eta(t, x) = x$ <u>if</u> $I(x) \notin [c - \bar{\varepsilon}, c + \bar{\varepsilon}]$

(2°) $\eta(1, A_{c+\varepsilon} \backslash \mathcal{O}) \subset A_{c-\varepsilon}$

(3°) <u>If</u> $K_c = \emptyset$, $\eta(1, A_{c+\varepsilon}) \subset A_{c-\varepsilon}$

PROOF. The proof can be found e.g. in [2] or [3]. We briefly sketch the main ideas for the special case of $E = \mathbb{R}^n$ and $I \in C^2(E, \mathbb{R})$. Consider the ordinary differential equation

$$\text{(1.2)} \qquad \begin{aligned} \frac{d\eta}{dt} &= -\chi(\eta) I'(\eta) \\ \eta(0,x) &= x \end{aligned}$$

where $\chi(z) \geq 0$ is a smooth real valued function which vanishes when $|I(z) - c| > \bar{\varepsilon}$ and is appropriately positive when $I(z)$ is near c. Then η trivially satisfies 1°. Moreover $I(\eta(t,x))$ does not increase along trajectories of (1.2). Indeed

$$\text{(1.3)} \qquad \frac{d}{dt} I(\eta(t,x)) = \left(I', \frac{d\eta}{dt}\right) = -\chi |I'|^2 \leq 0$$

This observation leads readily to 2° - 3°.

We can now describe a minimax principle. Let S denote some preferred family of subsets of E. For convenience we take each member of S to be compact. Define

$$\text{(1.4)} \qquad c = \inf_{K \in S} \max_{u \in K} I(u)$$

Clearly $c < \infty$ but it may well be the case that $c = -\infty$. Thus we assume $c > -\infty$. Given c and any pair $\bar{\varepsilon}$, $\mathcal{O}$ as in the statement of Lemma 1.1, there exists a corresponding η depending on $\bar{\varepsilon}$ and $\mathcal{O}$. Fix any such $\mathcal{O}$.

MINIMAX PRINCIPLE. _Suppose the above conditions on_ I _are satisfied and there is an_ $\bar{\varepsilon} > 0$ _such that_ $\eta(1,\cdot) : S \to S$. _Then_ c _is a critical value of_ I.

PROOF. If not, 3° of Lemma 1.1 obtains. Choose $K \in S$ such that

$$\max_{u \in K} I(u) \le c + \varepsilon$$

Since $\eta(1, K) \in S$,

$$\max_{u \in \eta(1,K)} I(u) \ge c \tag{1.5}$$

But by 3° of Lemma 1.1, $\eta(1,K) \subset A_{c-\varepsilon}$ so

$$\max_{u \in \eta(1,K)} I(u) \le c - \varepsilon ,$$

contrary to (1.5). This contradiction gives the result.

Next we give some applications. A very simple one is:

THEOREM 1.6. _Suppose_ $I \in C^1(E, \mathbb{R})$, _satisfies_ (PS), _and is bounded from below. Then_ $c = \inf_E I$ _is a critical value of_ I.

PROOF. Let $S = \{x \in E\}$ and $\bar{\varepsilon} > 0$. Then $\eta(1,\cdot) : S \to S$ and we trivially have

$$c = \inf_{K \in S} \max_{u \in K} I(u)$$

Next we give a geometrical application of the minimax principle due to Ambrosetti and the author [4] . Below $B_r = \{x \in E \mid \|x\| < r\}$.

THEOREM 1.7. _Let_ $I \in C^1(E, \mathbb{R})$ _and satisfy_ (PS). _Suppose_ $I(0) = 0$ _and_

(I_1) _There exist_ $\rho, \alpha > 0$ _such that_ $I > 0$ _in_ $B_\rho \setminus \{0\}$ _and_ $I \geq \alpha$ _on_ ∂B_ρ .

(I_2) _There is an_ $e \in E$, $e \neq 0$ _such that_ $I(e) = 0$.

Then I _has a positive critical value_.

PROOF. Set $\Gamma = \{g \in C([0,1],E) \mid g(0) = 0 , \ g(1) = e\}$ and $S = \{g([0,1]) \mid g \in \Gamma\}$. Thus S consists of curves joining 0 and e . Define

$$c = \inf_{K \in S} \max_{u \in K} I(u)$$

Since each curve K crosses ∂B_ρ , $\max_K I \geq \alpha > 0$. Hence $c \geq \alpha > 0$. We claim c is a critical value of I . By the minimax principle all we need do to verify this is to show that for some $\bar{\varepsilon} > 0$, $\eta(1,S) \subset S$. Choose $\bar{\varepsilon} < \frac{\alpha}{2}$. By the definition of S , it suffices to show $\eta(1,\cdot) : \Gamma \to \Gamma$. Since $\bar{\varepsilon} \leq c/2$, $\eta(1,x) = x$ if $I(x) \leq \frac{\alpha}{2}$ via 1° of Lemma 1.1. In particular $\eta(1,0) = 0$ and $\eta(1,e) = e$. Therefore $\eta(1,g(0)) = 0$ and $\eta(1,g(1)) = e$ if $g \in \Gamma$. Clearly $\eta(1,g) \in C([0,1],E)$. Hence $\eta(1,\cdot) : \Gamma \to \Gamma$ and the proof is complete.

REMARK 1.8. If more structure is imposed on I, a much stronger conclusion obtains. Namely if I is also even, i.e. $I(u) = I(-u)$, and (I_2) is replaced by the condition that for all finite dimensional subspaces $\tilde{E} \subset E$, there is an $R = R(\tilde{E}) > 0$ such that $I(u) \le 0$ for $u \in \tilde{E} \setminus B_{R(\tilde{E})}$, then I possess $\dim(E)$ critical points. (In particular infinitely many if E is infinite dimensional [4].)

As an application of Theorem 1.7, consider the boundary value problem.

$$(1.9) \qquad \begin{aligned} -\Delta u &= p(x,u) , & x &\in \Omega \\ u &= 0 , & x &\in \partial\Omega \end{aligned}$$

where Ω is a bounded domain in $\mathbb{R}^n$, $n \ge 3$, with a smooth boundary, p is e.g. smooth in its arguments, and satisfies

(p_1) $|p(x,z)| \le a_1 + a_2|z|^s$, $\quad s < \frac{n+2}{n-2}$.

(p_2) $p(x,z) = o(|z|)$ at $z = 0$.

(p_3) $0 < P(x,z) = \int_0^z p(x,t)\,dt \le \theta z p(x,z)$ for $|z| \ge M$ where $\theta \in (0,\frac{1}{2})$.

(When $n = 1, 2$, the growth condition (p_1) can be considerably relaxed.) Above and henceforth, a_1, a_2, etc. denote positive constants.

Note that (p_2) - (p_3) are superlinearity conditions for p at $z = 0$ and ∞. In particular, dividing the inequality in (p_3) by zP and integrating shows

$$P(x,z) \geq a_3 |z|^{\frac{1}{\theta}} - a_4$$

so

(1.10) $$|p(x,z)| \geq a_5 |z|^{\frac{1}{\theta} - 1} - a_6$$

for all $x \in \bar{\Omega}$, $z \in \mathbb{R}$.

Define

(1.11) $$I(u) = \tfrac{1}{2} \int_\Omega |\nabla u|^2 dx - \int_\Omega P(x,u)\, dx \quad .$$

Formally, critical points of I on $E = W_0^{1,2}(\Omega)$ (using the usual Sobolev space notation) are weak solutions of (1.9). Standard regularity theorems show weak solutions of (1.9) are smooth when p is smooth and hence are classical solutions of (1.9). We claim that I satisfies the hypotheses of Theorem 1.7 and therefore (1.9) possesses a nontrivial solution. We will sketch the proof. More details can be found in [4].

First (p_1) and a standard result imply I is defined on all of E and $I \in C^1(E, \mathbb{R})$. Clearly $I(0) = 0$. By (p_2) and the Poincaré and Sobolev inequalities, it is easy to verify that for each $\varepsilon > 0$, there is a $\delta > 0$ such that $\|u\| \leq \delta$ implies

(1.12) $$\left| \int_\Omega P(x,u)\, dx \right| \leq \varepsilon \|u\|_E^2$$

Since we can take $(\int_\Omega |\nabla u|^2 dx)^{\frac{1}{2}}$ as norm in E, the form of I then implies that I satisfies (I_1). Next observe that for any $u \in E$ with e.g. $\|u\|_E = 1$,

$$I(\alpha u) = \frac{\alpha^2}{2} \|u\|_E^2 - \int_\Omega P(x, \alpha u)\, dx$$

so I satisfies (I_2) via (1.10). Lastly to verify (PS), suppose

$|I(u_m)| \le K$ and $I'(u_m) \to 0$ as $m \to \infty$. Then for all large m,

$$(1.13) \qquad |I'(u_m)\varphi| \le \|\varphi\|_E$$

Choosing in particular $\varphi = u_m$ gives

$$(1.14) \qquad |\int (|\nabla u_m|^2 - p(x,u_m)\, u_m)\, dx| \le \|u_m\|_E .$$

Since

$$(1.15) \qquad K \ge I(u_m) = (\tfrac{1}{2} - \theta)\int_\Omega |\nabla u_m|^2\, dx + \theta \int_\Omega |\nabla u_m|^2\, dx$$

$$+ \int_{\{x\in\Omega : |u_m(x)| \le M\}} P(x,u_m)\,dx + \int_{\{x\in\Omega : |u_m(x)| > M\}} P(x,u_m)\,dx$$

$$\ge (\tfrac{1}{2} - \theta)\int_\Omega |\nabla u_m|^2\, dx + \theta \int_\Omega (|\nabla u_m|^2 - p(x,u_m)u_m)\, dx - a_7$$

$$\ge (\tfrac{1}{2} - \theta)\|u_m\|_E^2 - \theta\|u_m\|_E - a_7$$

(where we used (p_3)), it follows that (u_m) is bounded in E. Hence (u_m) has a weakly convergent subsequence. By (p_2), in operator form $I'(u_m) = u_m - \mathcal{P}(u_m)$ where $\mathcal{P}$ is compact. Hence $\mathcal{P}(u_m)$ has a convergent subsequence. Therefore $u_m = I'(u_m) + \mathcal{P}(u_m)$ has a convergent subsequence. Thus (PS) is verified and we have shown

THEOREM 1.16. _If_ p _satisfies_ $(p_1)-(p_3)$, (1.9) _possesses at least one nontrivial solution_.

REMARK 1.17. An additional argument shows (1.9) has a positive and a negative solution. (By a positive solution u, we mean $u > 0$ in Ω and satisfies (1.9).) If $p(x,z)$ is also odd in z, the result mentioned in Remark 1.8 shows (1.9) has infinitely many distinct solutions, indeed an unbounded sequence of solutions.

In Theorem 1.7, I has a known critical point to begin with, namely $u=0$. In our next abstract application, we study a situation where this is not the case.

THEOREM 1.18. Let $I \in C^1(E,\mathbb{R})$ and satisfy (PS). Suppose $E = E_1 \oplus E_2$ with E_1 finite dimensional. If there exist constants $b_1 < b_2$ and a neighborhood, Ω, of 0 in E_1 such that

$$(I_3) \quad I|_{E_2} \geq b_2$$

and

$$(I_4) \quad I|_{\partial\Omega} \leq b_1 \;,$$

then I has a critical point.

PROOF. Set $\Gamma = \{\chi \in C(\bar{\Omega},E) \mid \chi(u) = u \text{ if } u \in \partial\Omega\}$ and $S = \{\chi(\bar{\Omega}) \mid \chi \in \Gamma\}$. Define c as usual by (1.4). We claim c is a critical value of I. To see that, note first that $c \geq b_2$. Indeed if $\chi \in \Gamma$, $\chi = \varphi + \psi$ where $\varphi \in C(\bar{\Omega},E_1)$, $\psi \in C(\bar{\Omega},E_2)$, and $\varphi(u) = u$ if $u \in \partial\Omega$. Therefore there is a $\bar{z} \in \Omega$ such that $\varphi(\bar{z}) = 0$ and $\chi(\bar{z}) = \psi(\bar{z}) \in E_2$. By (I_3), $I(\bar{z}) \geq b_2$. It follows that $\max_{\chi(\bar{\Omega})} I \geq b_2$ and $c \geq b_2$. Next choose $0 < \bar{\varepsilon} < \frac{1}{2}(b_2 - b_1)$. This is possible via (I_4). If we show $\eta(1,\cdot)$ as given by Lemma 1.1 maps $S \to S$, then the minimax principle implies c is a critical value of I. As in the proof of Theorem 1.7, this will be verified if we show $\eta(1,\cdot) : \Gamma \to \Gamma$. Clearly $\eta(1,h) \in C(\bar{\Omega},E)$ if $h \in \Gamma$. Moreover if $x \in \partial\Omega$, $h(x) = x$ and $I(x) \leq b_1 < b_1 + \bar{\varepsilon} < b_2 - \bar{\varepsilon}$. Hence by 1° of Lemma 1.1, $\eta(1,x) = x$ and the proof is complete.

As an application of Theorem 1.18, we will obtain a generalization of a recent result of Ahmad, Lazer, and Paul [5]. Consider

$$(1.19) \qquad Lu \equiv -\sum_{i,j=1}^{n} (a_{ij}(x)u_{x_j})_{x_i} + c(x)u = f(x,u)\ , \ x \in \mathscr{D}$$

$$u = 0\ , \quad x \in \partial\mathscr{D}$$

where L is uniformly elliptic with smooth coefficients in $\mathscr{D}$, a bounded domain in $\mathbb{R}^n$ with a smooth boundary. We assume

(f_1) $f \in C(\Omega \times \mathbb{R}, \mathbb{R})$ and there are constants ε, $M > 0$ such that $|f(x,z)| \le \varepsilon|z| + M$ for all $x \in \bar{\Omega}$, $z \in \mathbb{R}$.

If $N(L) = \{0\}$, where $N(L)$ denotes the null space of L, then it is a trivial exercise to use (f_1) and the Schauder fixed-point theorem to get a solution of (1.19) provided that ε is sufficiently small. Thus (1.19) is of interest when $N(L)$ is non-trivial. The ellipticity of L then implies

(L_1) $N(L)$ is finite dimensional.

Let $\varphi_1, \ldots, \varphi_p$ span $N(L)$. We cannot expect to be able to solve (1.19) without further restrictions on f. This is even true if f is independent of u. Two sufficient conditions for existence are

(f_2^+) ((f_2^-)) If $F(x,z) = \int_0^z f(x,t)\,dt$, then $\int_\Omega F(x,\varphi(x))\,dx \to +\infty$ ($\to -\infty$) as $|\varphi| \to \infty$ for all $\varphi \in N(L)$.

Indeed we have

THEOREM 1.20. *If* (f_1), (L_1), *and* (f_2^+) (or (f_2^-)) *are satisfied, and* ε *is sufficiently small,* (1.19) *possesses at least one weak solution.*

PROOF. Let $I(u) = \frac{1}{2}\int_\Omega \left(\sum_{i,j=1}^{n} a_{ij}(x) u_{x_i} u_{x_j} + c(x)u^2 \right) dx - \int_\Omega F(x,u)\,dx$

We will show I has a critical point in $E = W_0^{1,2}(\Omega)$ from which the theorem follows. We claim I satisfies the hypotheses of Theorem 1.18. As in Theorem 1.16, (f_1) implies $I \in C^1(E, \mathbb{R})$. The definition of the subspaces E_1 and E_2 depends on which of (f_2^+), (f_2^-) we assume. We will only carry out the details for the (f_2^+) case; the remaining case is similar. Let N^+ (resp. N^-) denote the subspace of E on which L is positive (resp. negative) definite. Then N^+, N^-, and $N(L)$ are orthogonal with respect to the L^2 inner product and $E = N^+ \oplus N^- \oplus N(L)$. Set $E_1 = N^- \oplus N(L)$ and $E_2 = N^+$. For $u \in E_2$,

$$(1.21) \qquad I(u) \geq \beta \|u\|_E^2 - \int_\Omega F(x,u)\,dx \geq \beta \|u\|_E^2 - \left(\frac{\varepsilon}{2}\int_\Omega u^2 dx + M\int_\Omega |u|\,dx\right)$$

from which it follows that I is bounded from below on E_2 provided that ε is sufficiently small. Therefore (I_3) is satisfied.

For $u \in E_1$, $u = Pu + P^+ u$ where P, $P^\pm$ denote the L^2 orthogonal projectors of E onto $N(L)$, $N^\pm$. Moreover for such u

$$(1.22)\qquad I(u) \le -\gamma \|P^- u\|_E^2 - \int_\Omega [F(x,Pu) + F(x,Pu+P^- u) - F(x,Pu)]\,dx$$
$$\le -\gamma \|P^- u\|_E^2 - \int_\Omega F(x,Pu)\,dx + \frac{\varepsilon}{2}\int_\Omega u^2 dx + M\int_\Omega |u|\,dx$$

which implies $I(u) \to -\infty$ as $\|u\|_E \to \infty$. Hence choosing Ω to be a large ball in E_1, we can satisfy (I_4). Lastly to verify (PS), suppose

$$(1.23)\qquad |I(u_m)| \le K \text{ and } I'(u_m) \to 0 \text{ as } m \to \infty .$$

Consider

$$I'(u)\varphi = \int_\Omega \Big[\sum_{i,j=1}^n a_{ij}(x)\, u_{x_j} \varphi_{x_i} + c(x)u\varphi - f(x,u)\varphi \Big] dx$$

with $\varphi = P^\pm u_m$. Then (f_1) implies $|I'(u)\, P^\pm u|$ grows like $\|P^\pm u\|_E^2$ as $\|P^\pm u\|_E \to \infty$. Since $I'(u_m) \to 0$ as $m \to \infty$, $\|P^\pm u_m\|_E$ must be bounded. If $\|P u_m\|_E$ were unbounded, (1.22) would imply $I(u_m) \to -\infty$ contrary to (1.23). Thus (u_m) is bounded in E and hence possesses a weakly convergent subsequence. Clearly $(P + P^-)u_m$ converges strongly along this subsequence. The argument used to verify (PS) in Theorem 1.16 further shows $(P^+ u_m)$ has a convergent subsequence.

REMARK 1.24. A closer examination of the above proof shows we need only require $\varepsilon < s$, the smallest element in magnitude in $\sigma(L)\setminus\{0\}$. If e.g. $\varepsilon = 0$ in (f_1), $(f_2^{\pm})$ is satisfied if $F(x,z) \to \pm\infty$ as $|z| \to \infty$, uniformly in x. Further applications of the ideas used in Theorem 1.18 can be found in [6].

§2. MULTIPLE CRITICAL POINTS

In this section we will give some examples showing how additional structure can be used to obtain multiple critical points of I. Our goal is to prove a result due to Clark [2] which is perhaps the simplest of this type. Before stating it, some additional terminology is needed. Let $\Sigma = \{A \subset E \setminus \{0\} \mid A$ is closed (in E) and symmetric with respect to $0\}$. Thus $A \in \Sigma$ and $x \in A$ implies $-x \in A$. We define a mapping from Σ to $\mathbb{N}$ as follows: If $A \in \Sigma$, $\gamma(A)$, the <u>genus</u> of A is the least integer n such that there is an odd map $\varphi \in C(A, \mathbb{R}^n \setminus \{0\})$. If $A = \emptyset$, we define $\gamma(A) = 0$. As simple examples, if $x \in E \setminus \{0\}$, $\gamma(\{x, -x\}) = 1$ while if $A \subset E$ is connected, $\gamma(A) > 1$. A rich source of sets of nontrivial genus is provided by the following result:

LEMMA 2.1. <u>If</u> h <u>is an odd homeomorphism from</u> S^{n-1} <u>to</u> E, <u>then</u> $\gamma(h(S^{n-1})) = n$.

PROOF. The proof is a simple consequence of the Borsuk-Ulam Theorem and can be found e.g. in [3] or [7].

The properties of genus we require are contained in the following lemma:

LEMMA 2.2. *Let* $A, B \in \Sigma$.

(1°) *If there is an odd* $f \in C(A,B)$, *then* $\gamma(A) \le \gamma(B)$

(2°) *If* $A \subset B$, $\gamma(A) \le \gamma(B)$

(3°) $\gamma(A \cup B) \le \gamma(A) + \gamma(B)$

(4°) *If* $\gamma(B) < \infty$, $\gamma(\overline{A \setminus B}) \ge \gamma(A) - \gamma(B)$

(5°) *If* A *is compact,* $\gamma(A) < \infty$ *and there exists a* $\delta > 0$ *such that if* $N_\delta(A) \equiv \{u \in E : \|u - A\| \le \delta\}$, *then* $\gamma(N_\delta(A)) = \gamma(A)$.

PROOF. The proofs are all elementary and will be omitted.

With these preliminaries in hand, we can state Clark's result. Let $\Sigma_j = \{A \in \Sigma \mid A$ is compact and $\gamma(A) \ge j\}$.

THEOREM 2.3. *Let* $I \in C^1(E, \mathbb{R})$ *be even and satisfy* $I(0) = 0$ *and* (PS). *Define*

$$c_j = \inf_{A \in \Sigma_j} \max_{u \in A} I(u). \tag{2.4}$$

Suppose $-\infty < c_j < 0$, $1 \le j \le k$. *Then* I *possesses at least* k *distinct pairs of critical points*.

PROOF. First we show c_j is a critical value of I, $1 \le j \le k$. To do this we need not require that $I(0) = 0$ or that $c_j < 0$. We invoke the minimax principle with $S = \Sigma_j$ and any $\bar{\varepsilon}$, $\mathcal{O}$ which is legitimate if $\eta(1,\cdot) : \Sigma_j \to \Sigma_j$. But this is evident from 1° of Lemma 2.2 with $B = \eta(1,A)$, $A \in \Sigma_j$.

It remains to show we have k distinct critical points. Since $I(0) = 0$ and $c_j < 0$, $1 \le j \le k$, $K_{c_j} \in \Sigma$. Each distinct critical value c_j gives us a corresponding pair of critical points. Thus all we need verify is that we have the correct number of critical points. Suppose that $c_{j+1} = \dots = c_{j+p} \equiv c$ for some $p > 1$. Then we will show $\gamma(K_c) \ge p$. If K_c were a finite point set, by the definition of genus, $\gamma(K_c)$ would equal one. Hence $\gamma(K_c) > 1$ implies K_c contains infinitely many points so we obtain more than enough critical points for this case. To show that $\gamma(K_c) \ge p$, assume to the contrary that $\gamma(K_c) \le p-1$. Then by 5° of Lemma 2.2, there is a $\delta > 0$ such that $\gamma(N_\delta(K_c)) \le p-1$. Choose $\mathcal{O} = \operatorname{int} N_\delta(K_c)$ in Lemma 1.1 and any $\bar{\epsilon} > 0$. Then with ϵ as in that lemma, choose $A \in \Sigma$ such that $\gamma(A) \ge j+p$ and $\max_A I \le c+\epsilon$. By 4° of Lemma 2.2, $\gamma(A \backslash N_\delta(K_c)) \ge j+p-(p-1) = j+1$. Letting $B = \eta(1, A\backslash N_\delta(K_c))$, by 1° of Lemma 2.2, $\gamma(B) \ge j+1$. Thus B is an admissible set for computing $c = c_{j+1}$. But since $B \subset A_{c-\epsilon}$ by 2° of Lemma 1.1, $\max_B I \le c - \epsilon$ contrary to the definition of c_{j+1}. The proof is complete.

REMARK 2.5. A similar argument can be used to prove the result of Ljusternick - Schnirelman mentioned in the introduction. Namely by taking $E = \mathbb{R}^n$, working with subsets of Σ which lie in S^{n-1}, and using a version of Lemma 1.1 appropriate to S^{n-1}, the above proof goes through virtually unchanged.

As an application of Theorem 2.3, consider

$$(2.6) \qquad -\Delta u = \lambda(u - p(x,u)) \qquad x \in \Omega$$
$$u = 0 \quad , \quad x \in \partial\Omega$$

where Ω is a bounded domain in $\mathbb{R}^n$ with a smooth boundary and p satisfies (p_2) and

(p_4) There is a $\bar{z} > 0$ such that $p(x,\bar{z}) > \bar{z}$.

(p_5) $p(x,z)$ is even in z .

If we drop the p term in (2.6), we obtain a corresponding linear eigenvalue problem

$$(2.7) \qquad -\Delta v = \lambda v \qquad x \in \Omega$$
$$v = 0 \qquad x \in \partial\Omega$$

This problem possesses an increasing sequence of eigenvalues λ_n of finite multiplicity with $0 < \lambda_1 < \lambda_2 \leq \cdots \leq \lambda_m \to \infty$ as $m \to \infty$. As a consequence of Theorem 2.3, we can show

THEOREM 2.8. _Let_ p _satisfy_ (p_2) , (p_4) , (p_5) _and let_ $\lambda > \lambda_k$. _Then_ (2.6) _possesses at least_ k _distinct pairs of nontrivial solutions_.

PROOF. Define $f(x,z) = z - p(x,z)$ if $|z| \leq \bar{z}$; $f(x,z) = \bar{z} - p(x,\bar{z})$ if $z > \bar{z}$, and $f(x,z) = -\bar{z} - p(x,-\bar{z})$ if $z < -\bar{z}$. Consider

$$(2.9) \qquad -\Delta u = \lambda f(x,u) \quad , \quad x \in \Omega$$
$$u = 0 \quad , \quad x \in \partial\Omega$$

It is an easy exercise using the maximum principle to show that any solution of (2.9) with $\lambda \geq 0$ satisfies $|u(x)| \leq \bar{z}$. Hence (λ, u) is a solution of (2.6). Thus we focus our attention on (2.9) which has a bounded right hand side.

Consider

$$(2.10) \qquad I(u) = \frac{1}{2}\int_\Omega |\nabla u|^2 dx - \lambda \int_\Omega F(x,u)\,dx$$

where $F(x,z) = \int_0^z f(x,s)\,ds$. Thus $I(0) = 0$. As usual we obtain the desired number of solutions of (2.9) as critical points of I. The boundedness of f implies $I \in C^1(E,\mathbb{R})$ with $E = W_0^{1,2}(\Omega)$. Condition (p_5) implies I is even and by the boundedness of f, I is bounded from below. Hence c_j as defined via (2.4) is finite for all $j \in \mathbb{N}$. To complete the proof, we must show that I satisfies (PS) and that $c_j < 0$, $1 \leq j \leq k$. To verify the former condition, note that if $I(u_m)$ is bounded, then (u_m) must be bounded. Hence a subsequence of (u_m) converges as in the proof of Theorem 1.16. Next to see that $c_j < 0$ if $j \leq k$, let v_j denote an eigenfunction of (2.7) with $\lambda = \lambda_j$ normalized so that $\|v_j\|_E = 1$. Consider

$$T = \left\{ \sum_{j=1}^{k} \alpha_j v_j \mid \sum_{j=1}^{k} \alpha_j^2 = \rho^2 \right\}$$

By Lemma 2.1, $\gamma(T) = k$. If ρ is small enough and $u \in T$,

$$I(u) = \frac{1}{2}\rho^2 - \frac{\lambda}{2}\int_\Omega \sum_{j=1}^{k} \alpha_j^2 v_j^2 \,dx + o(\rho^2)$$

$$= \frac{1}{2}\left(\rho^2 - \lambda \sum_{j=1}^{k} \frac{\alpha_j^2}{\lambda_j}\right) + o(\rho^2) < \frac{1}{2}\rho^2\left(1 - \frac{\lambda}{\lambda_k}\right) < 0$$

Hence $c_j < 0$ and the proof is complete.

REMARK 2.11. For other results in this direction see e.g. [3]

REFERENCES

[1] Ljusternick, L.A. and L.G. Schnirelman, Topological Methods in the Calculus of Variations, Hermann, Paris, 1934.

[2] Clark, D.C., A variant of the Ljusternick-Schnirelman theory, Indiana Univ. Math. J. 22 (1972), 65-74.

[3] Rabinowitz, P.H., Variational methods for non-linear eigenvalue problems, Proc. Sym. on Eigenvalues of Nonlinear Problems, Edizioni Cremonese, Rome, 1974, 141-195.

[4] Ambrosetti, A. and P.H. Rabinowitz, Dual variational methods in critical point theory and applications, J. Functional Anal., 14, (1973), 349-381.

[5] Ahmad, S., A.C. Lazer, and J.L. Paul, Elementary critical point theory and perturbations of elliptic boundary value problems at resonance, to appear Ind. Univ. Math. J.

[6] Rabinowitz, P.H., Some minimax theorems and applications to non-linear partial differential equations, to appear.

[7] Coffman, C.V., A minimax principle for a class of nonlinear integral equations, J. Analyse Math. 22 (1969), 391-419.

THE EXISTENCE OF PERIODIC WATER WAVES

John Norbury
Department of Mathematics
University College, London

1. Consider the two-dimensional flow of an ideal liquid in the $x > 0$ direction over the flat bottom $y = 0$. As shown in figure 1 gravity acts in the $-y$ direction, and the flow is bounded from above by the free boundary $\Gamma = \{(x,y) \mid y = h(x) > 0$ and $h(x) = h(x + 2L)$ for all $x\}$. The periodicity of Γ in x forces the whole flow to be periodic in x, and when $h(x)$ is non-constant we call this a wave flow. We shall prove that there exist wave flows, and that the wave solutions bifurcate from the trivial solutions, the uniform streams for which $h(x) \equiv$ constant.

More precisely, our problem (P) is to find a free boundary Γ and a complex velocity $u-iv$ (where u and v are, respectively, the x and y velocity components), such that $u-iv$ is an analytic function of the complex position $x+iy$ and the following boundary conditions are satisfied:

$$u+iv \text{ is tangential to } \Gamma \text{ when } (x,y) \in \Gamma , \tag{1}$$

$$v = 0 \text{ when } y = 0 \text{ for all } x , \tag{2}$$

and

$$\tfrac{1}{2}(u^2+v^2) + gy = R = \text{constant when } (x,y) \in \Gamma . \tag{3}$$

(Equation (3) is known as Bernoulli's equation, and it follows from the constancy of the pressure along Γ.) Further, we look only for

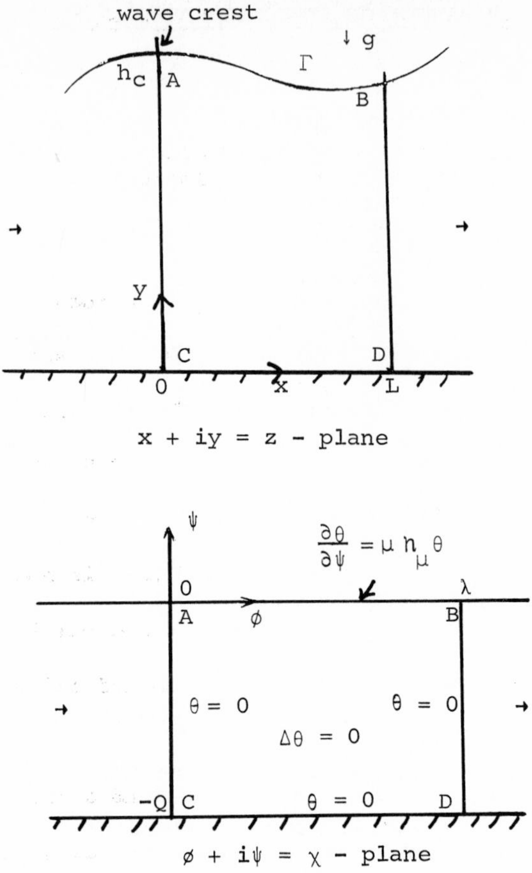

FIGURE 1

An ideal liquid flows over a flat bottom with gravity acting in the -y direction. The unknown flow domain is mapped conformally into the strip, and θ is the angle that the flow velocity makes with the x axis.

Γ and u-iv which are periodic in x with period $2L$, and we normalise (P) by taking the flux of fluid to be Q and the velocity at the crest of the wave to be $q_c > 0$.

A trivial solution of (P) is given by $u-iv \equiv q_c > 0$ and $h(x) \equiv Q/q_c$. We seek a "branch" of wave solutions of (P) which exist for positive values of the parameter $\mu = g/q_c^3$, and which bifurcate from the trivial solutions when $\mu = \mu_0$ (figure 2). We define $\theta = \arg(u-iv)$ (θ is the angle the velocity makes with the x axis), and $\|\theta\|$ to be the maximum of $|\theta|$ in the flow. (The maximum of $|\theta|$ must occur on Γ.) We will establish, in Theorem 4, the existence of an unbounded component C of wave solutions of (P) with $\mu > \mu_0$ and $\|\theta\| < \pi/2$ on the component. Further, $\mu \to \mu_0$ as $\|\theta\| \to 0$.

There is an extensive literature on the local bifurcation theory for (P) (see Hyers), but the only previous global results for (P) are those of Krasovskii. He proved that for each $\|\theta\| \in (0,\pi/6)$ there is a wave solution with $\mu > 0$. Since our C includes Krasovskii's solutions it follows that $\rho = \sup_C \|\theta\| \geq \pi/6$. Longuet-Higgins and Fox give numerical evidence that $\rho > \pi/6$.

2. The unknown boundary Γ and the non-linear boundary conditions make the solution of problem (P) difficult. We thus seek new independent and dependent variables so as to reformulate (P) as a mixed boundary value problem in a standard domain. First we conformally map the flow domain onto the strip $E = \{-Q < \psi < 0, -\infty < \phi < \infty\}$ so that Γ is mapped to $\{\psi=0\}$, $\{y=0\}$ is mapped to

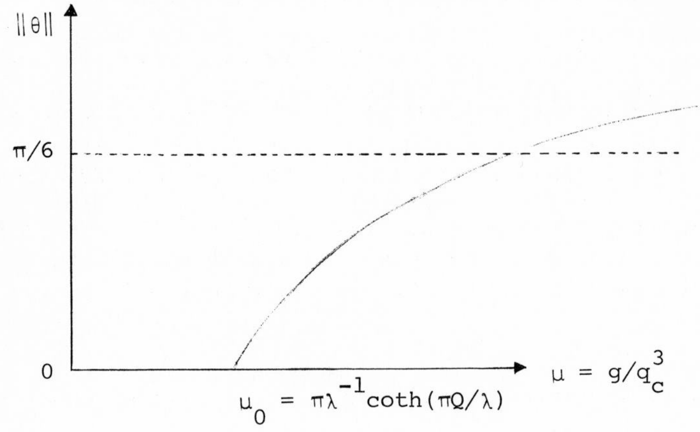

FIGURE 2

A possible bifurcation diagram. It is proved that the component C of solutions (θ,μ) emanating from $(0,\mu_0)$ is unbounded, and that $\|\theta\| < \pi/2$ and $\mu > \mu_0$ on C .

$\{\psi = -Q\}$, and the crest of a wave is mapped to $(0,0)$. The conformal mapping from $x + iy$ to $\phi + i\psi$ is defined by means of the relations $u = \phi_x = \psi_y$ and $v = \phi_y = -\psi_x$ which hold in the flow domain (figure 1). (The stream function $\psi(x,y)$ thus defined is constant along the lines of flow, and so is constant along the flow boundaries.)

Next, we define $\tau + i\theta = \ell n(u-iv)$, where $\tau + i\theta$ is real and positive when $u - iv$ is real and positive. (The definitions make sense since $u > 0$ throughout the flow. This follows from use of the maximum principle and the a priori inequalities $\|\theta\| < \pi/2$ derived in Theorem 3.) Further, since $x + iy$ is analytic in $\phi + i\psi$, $\tau + i\theta$ is analytic in $\phi + i\psi$.

With these variables we can state a reformulated problem (P^1) involving only the function $\theta(\phi,\psi)$: find $\mu > 0$ and $\theta(\phi,\psi)$ harmonic in E , periodic in ϕ with period 2λ and odd in ϕ , satisfying the boundary conditions

$$\theta(0,\psi) = \theta(\lambda,\psi) = 0 = \theta(\phi,-Q) \tag{4}$$

for $-\infty < \phi < \infty$ and $-Q \leq \psi \leq 0$, and

$$\frac{\partial\theta}{\partial\psi} = \mu h_\mu \theta \equiv \mu \sin\theta \, \{1 + 3\mu \int_0^\phi \sin\theta\,(\hat{\phi},0)\, d\hat{\phi}\}^{-1} \tag{5}$$

on $\psi = 0$ for $-\infty < \phi < \infty$. Equation (5) follows from:

(i) differentiating (3) with respect to ϕ along Γ , using $u^2 + v^2 = \exp(2\tau)$ and $\tau_\phi = \theta_\psi$ in E , and, defining s as arc length along Γ , we find

$$e^{2\tau}\theta_{\psi} = -gy_{\phi} = -gy_s s_{\phi}$$

on $\psi = 0$;

(ii) noting that $-y_s = \sin\theta$ and, along streamlines, $\phi_s = |d(\phi+i\psi)/d(x+iy)| = \{u^2+v^2\}^{\frac{1}{2}} = e^{\tau}$, we have

$$\theta_{\psi} = g \sin\theta \cdot \exp(-3\tau)$$

on $\psi = 0$; and

(iii) integrating $(\exp 3\tau)_{\phi} = 3\exp(3\tau)\tau_{\phi} = 3\, g \sin\theta$ along $\psi = 0$ from $\phi = 0$ we obtain

$$\exp(3\tau) = q_c^3\{1 + 3gq_c^{-3}\int_0^{\phi} \sin\theta(\hat{\phi},0)\,d\hat{\phi}\}$$

on $\psi = 0$.

A solution (θ,μ) of (P^1) leads to a solution of (P) by means of $u-iv = \exp(\tau+i\theta)$, where τ is the normalised harmonic conjugate of θ , and $x+iy = \int (u-iv)^{-1} d(\phi+i\psi)$.

3. Although the problem (P^1) has a non-linear boundary condition (5) that is integro-differential we consider (P^1) as if it was a standard mixed boundary value problem. We introduce the Green's function

$$G(\phi,\hat{\phi},\psi) = \frac{2}{\pi}\sum_{n=1}^{\infty} \frac{\sinh(n\pi(\psi+Q)/\lambda)}{n\cosh(n\pi Q/\lambda)} \sin(n\pi\phi/\lambda)\sin(n\pi\hat{\phi}/\lambda) \qquad (6)$$

so that

$$\theta(\phi,\psi) = \int_0^{\lambda} G(\phi,\hat{\phi},\psi)\mu h_{\mu}\theta(\hat{\phi},0)\,d\hat{\phi} \tag{7}$$

is the solution of (P^1) in E.

We define $\theta(\phi) = \theta(\phi,0)$. Then, on letting $\psi \to 0$ in (7), we have

$$\theta(\phi) = \mu\int_0^{\lambda} G(\phi,\hat{\phi},0)h_{\mu}\theta(\hat{\phi})\,d\hat{\phi}\,, \tag{8}$$

$0 \leq \phi \leq \lambda$; that is, for $\mu \geq 0$ and $0 \leq \phi \leq \lambda$,

$$\theta = \mu\mathcal{G}h_{\mu}\theta \equiv \mathcal{G}(\theta,\mu)\,. \tag{9}$$

If we can find solutions θ of the non-linear integral equation (9) then (7) gives us a solution of (P^1), and hence we can construct a solution of the original problem (P). Note that $\theta = 0$ is a solution of (9) for all $\mu \geq 0$, and is the trivial solution corresponding to the uniform streams.

We now proceed to find solutions $\theta \neq 0$ of equation (9) in a set X in the Banach space B of continuous functions defined for $0 \leq \phi \leq \lambda$ with the norm $\|\cdot\| = \max_{0\leq\phi\leq\lambda} |\cdot(\phi)|$ and such that the functions vanish at $\phi = 0$ and $\phi = \lambda$. In fact, since whenever $\pi \geq \theta \geq 0$ and $\mu \geq 0$ we have $\sin\theta \geq 0$ and $h_{\mu}\theta \geq 0$, and, since $G(\phi,\hat{\phi},\psi) \geq 0$, $\mathcal{G}(\theta,\mu) \geq 0$, we will use the following positive operator result, and take X to be the intersection of the ball of radius $\pi/2$ in B with K, the cone of functions in B that are pointwise positive for $0 \leq \phi \leq \lambda$.

THEOREM 1. (Dancer - see Amann, section 18.) Suppose that $\mathfrak{F} : K \times R_+ \to K$ is completely continuous and that $\mathfrak{F}(\cdot,0) = 0 = \mathfrak{F}(0,\cdot)$. Suppose that

$$\mathfrak{F}(\theta,\mu) = \mu\mathfrak{L}\theta + \mathfrak{R}(\theta,\mu) \quad \text{for} \quad (\theta,\mu) \in K \times R_+ ,$$

where $\mathfrak{L}$ is a completely continuous linear operator and $\mathfrak{R} : K \times R_+ \to K$ is such that $\|\mathfrak{R}(\theta,\mu)\| = o(\|\theta\|)$ as $\|\theta\| \to 0$ uniformly in μ in some neighborhood of $\tilde{\mu} > 0$. Suppose that there exists a unique $(\theta_0,\mu_0) \in K \times R_+$ such that $\theta_0 = \mu_0\mathfrak{L}\theta_0$ with $\|\theta_0\| = 1$. Define $\Sigma^+ = \{(\theta,\mu) \in K \times R_+ \,|\, \theta = \mathfrak{F}(\theta,\mu),\ \theta \neq 0\} \cup \{(0,\mu_0)\}$.

Then bifurcation from the line of trivial solutions occurs, and Σ^+ contains an unbounded sub-continuum C emanating from $(0,\mu_0)$.

Further, if μ_0 is simple and $\mu\mathfrak{L}\theta > \theta$ for $(\theta,\mu) \in K \times R_+$, then $\mu \geq \mu_0$.

Since the operator $\mathfrak{h}_\mu$ may cease to be positive (or even defined) when $\theta > \pi$, we modify $\mathfrak{h}_\mu$ so that we may apply Theorem 1. Define

$$\mathfrak{m}\,\theta = \begin{Bmatrix} 0 & \text{for} & \theta < 0 \\ \theta & \text{for} & 0 \leq \theta \leq \pi \\ \pi & \text{for} & \pi < \theta \end{Bmatrix} \tag{10}$$

and

$$\tilde{\mathfrak{a}}(\theta,\mu) \equiv \mu\mathfrak{G}\tilde{\mathfrak{h}}_\mu\theta \equiv \mu\mathfrak{G}\mathfrak{h}_\mu\mathfrak{m}\theta \quad . \tag{11}$$

Then we have

THEOREM 2. <u>The equation</u> $\theta = \tilde{\mathfrak{G}}(\theta,\mu)$ <u>has an unbounded subcontinuum</u> C <u>of solutions</u> $(\theta,\mu) \in K \times R_+$ <u>emanating from</u> $(0,\mu_0 = \pi\lambda^{-1}\coth(\pi Q\lambda^{-1}))$ <u>with</u> $\mu > \mu_0$ <u>for</u> $\theta \neq 0$.

PROOF. (a) $\tilde{\mathfrak{G}}$ is positive since $\sin \mathfrak{m}\theta \geq 0$ and $G(\phi,\hat{\phi},0) \geq 0$.

(b) $\tilde{\mathfrak{G}}(\theta,\mu) = \mu\mathcal{G}\theta + \mathfrak{R}(\theta,\mu)$ for $(\theta,\mu) \in K \times R_+$, with $\|\mathfrak{R}(\theta,\mu)\| = o(\|\theta\|)$ as $\|\theta\| \to 0$ uniformly in μ. $\mathcal{G}$ is a completely continuous linear operator in B of a standard type, and, since $\tilde{\mathfrak{n}}_\mu$ is a bounded continuous operator in $K \times R_+$, $\tilde{\mathfrak{G}} : K \times R_+ \to K$ is completely continuous.

(c) The eigenfunctions of $\mathcal{G}$ in B are $\sin(n\pi\phi/\lambda)$, for $n = 1, 2, 3, \ldots$, with characteristic values $n\pi\lambda^{-1}\coth(n\pi Q\lambda^{-1})$. Thus the only eigenfunction of $\mathcal{G}$ in K is $\sin(\pi\phi/\lambda)$ with eigenvalue $\mu_0 = \pi\lambda^{-1}\coth(\pi Q\lambda^{-1})$. (Note that $B[0,\lambda] \subset L_2(0,\lambda)$ and $G(\phi,\hat{\phi},0)$ is symmetric.)

(d) If $\theta \neq 0$ and $(\theta,\mu) \in K \times R_+$ then $\tilde{\mathfrak{n}}_\mu\theta < \sin \mathfrak{m}\theta < \theta$. Thus we apply Theorem 1 with $\mathfrak{F} \equiv \tilde{\mathfrak{G}}$ to obtain Theorem 2. The fact that $\mu > \mu_0$ for $\theta \neq 0$ follows from the variational characterisation of μ_0 in $L_2(0,\lambda)$.

4. If $(\theta,\mu) \in C$ and $\|\theta\| < \pi$ then $\theta = \tilde{\mathfrak{G}}(\theta,\mu) = \mathfrak{G}(\theta,\mu)$, and so we have found solutions of equation (9) . We prove in Theorem 3 that all solutions of (P) have $\|\theta\| < \pi/2$. For $\|\theta\| < \pi/2$ a solution of (P^1) (obtained from equation (7) on using $(\theta,\mu) \in C$) leads to a solution of (P) (after an integration to obtain $\phi(x,y)$ and $\psi(x,y)$) . Thus for $(\theta,\mu) \in C$ with $\|\theta\| < \pi/2$ we can construct the corresponding wave solution of the original problem (P) .

Then, using Theorem 3, we show in Theorem 4 that all $(\theta,\mu) \in C$ must have $\|\theta\| < \pi/2$, and thus yield wave solutions of (P).

THEOREM 3. Any wave flow satisfying problem (P) must have $\|\theta\| < \pi/2$.

PROOF. Define the pressure by $p = R - \frac{1}{2}(u^2+v^2) - gy$, where $R = \frac{1}{2}q_c^2 + gh_c$ and h_c is the height of the crest of the waves. Then $p_{xx} + p_{yy} \le 0$ in the flow. Further, $p_x = 0$ at $x = 0$ and at $x = L$ for $y \ge 0$ and $p = 0$ for $(x,y) \in \Gamma$ by equation (3). Finally $p > 0$ on $y = 0$ since $p_y < 0$ on $x = 0$.

The Hopf maximum principle then implies (i) $p > 0$ in the flow and (ii) the outward normal derivative of p on Γ, where $p = 0$, is strictly negative. For contradiction assume there is a point $z \in \Gamma$ where $\theta = \pi/2$. Then $u(z) = 0$ and $v(z) < 0$. The outward normal derivative is $p_x < 0$. But, on using $u_y = v_x$, we have $p_x = -uu_x - vv_x = -vu_y$. Since we may choose z to be the point with maximum y such that $\theta(z) = \pi/2$ we must have $u_y(z) \ge 0$, and so $p_x \ge 0$, a contradiction. This proves the theorem, since the case $\theta = -\pi/2$ is handled similarly.

THEOREM 4. If $(\theta,\mu) \in C$ then $\|\theta\| < \pi/2$, so that (θ,μ) is a solution of equation (9) and $\theta(\phi,\psi)$ defined by equation (7) is a solution of problem (P^1). Further, a corresponding wave solution of problem (P) can be found with $\|\theta\| < \pi/2$ for each $\mu \in (\mu_0,\infty)$.

PROOF. Assume, for contradiction, that there is a $(\theta,\mu) \in C$ with $\|\theta\| = \pi/2$. Then, since $\tilde{\mathcal{C}}$ is compact in K, we may choose, for $n = 1, 2, 3, \ldots$, $(\theta_n,\mu_n) \in C$ and $(\theta_*,\mu_*) \in C$ such that $\|\theta_n\| \leq \pi/2$, $\|\theta_*\| = \pi/2$ and $(\theta_n,\mu_n) \to (\theta_*,\mu_*)$ in $K \times R_+$. But, since each (θ_n,μ_n) has a corresponding wave solution of (P), we may construct a wave solution of (P) corresponding to (θ_*, μ_*) with $\|\theta_*\| = \pi/2$. This contradicts Theorem 3, and hence all $(\theta,\mu) \in C$ have $\|\theta\| < \pi/2$. Thus the theorem follows from Theorem 2, and the observation that $\theta(\phi,\psi)$ defined by (7) satisfies $\theta_\psi = \mu h_\mu \theta$ on $\psi = 0$.

Since Krasovskii claims to have a connected set of solutions of (P) emanating from $(0,\mu_0)$ in our notation, the waves corresponding to C must include Krasovskii's solutions. Hence $\sup_C \|\theta\| \geq \pi/6$. The $\lim_{\mu\to\infty} \theta(\phi)$ is thought to yield the (conjectured) Stoke's wave of greatest height, for which $\theta(\phi)$ jumps by $\pi/3$ at $\phi = 0$ where $q_c = 0$. (In contrast our waves have Γ analytic.) Does $\|\lim_{\mu\to\infty} \theta(\phi)\| = \pi/6$ even though $\lim_{\mu\to\infty}\|\theta\| > \pi/6$? See also Keady and Norbury.

REFERENCES

Amman, H. Fixed point equations and nonlinear eigenvalue problems in ordered Banach spaces. SIAM Rev. 18 (1976), 620.

Dancer, E.N. Global solution branches for positive mappings, Arch. Rat. Mech. Anal. 52 (1973), 181.

Hyers, D.H. Some nonlinear integral equations in hydrodynamics, Nonlinear integral equations (Anselone, ed., University of Wisconsin Press, 1965).

Keady, G. and Norbury, J. On the existence theory for irrotational water waves. Math. Proc. Camb. Phil. Soc. (1977).

Krasovskii, Yu. P. On the theory of steady state waves of large amplitude. U.S.S.R. Comp. Maths. and Math. Phys. 1 (1961), 996.

Longuet-Higgins, M.S. and Fox, M.J.H. Theory of the almost highest wave: the inner solution. J. Fluid Mech. 80 (1977), 721.

AN APPLICATION OF THE NASH-MOSER THEOREM TO A FREE BOUNDARY PROBLEM

David G. Schaeffer *
Mathematics Research Center
University of Wisconsin
Madison, Wisconsin

This note discusses an application of the Nash-Moser implicit function theorem to a free boundary problem. This theorem, originally introduced in a special case by Nash [5] and later abstracted and simplified by Moser [4], is a powerful generalization of the usual implicit function theorem in a Banach space. It enables one to solve certain non-linear partial differential equations, say, for example, $F(x,u, \text{grad } u) = 0$, with the unpleasant property that both the differential dF and its inverse are unbounded linear operators. This is an extreme example of the famous "loss of derivations" phenomenon. Very strong control of the differentiability of F with respect to u is required in exchange.

Several applications of this theorem to free boundary problems have appeared ([3], [6], [7], [8]). Of these, three involve elliptic problems while [8] involves a hyperbolic problem. Here we discuss only [8]. We introduce the problem, briefly review the derivation of the governing equations from gas dynamics, state the principal theorem about the existence of a solution, and summarize the main points of the proof. Complete proofs are given in [8], but they are somewhat lengthy; here we try to avoid technical complications in favor of a (we hope) clearer exposition.

Nearly everyone has seen pictures of the shock wave generated by a supersonic projectile (Figure 1 is a poor substitute for such a picture).

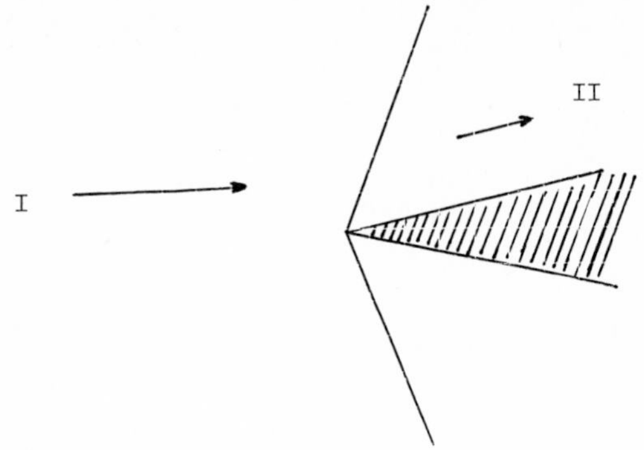

Figure 1

Let us treat this problem in the approximation of neglecting viscosity and heat conduction. We suppose that the flow is two-dimensional. We may pass to the coordinate system in which the projectile is at rest, so we are looking for a steady state flow. If the surface of the wedge is perfectly straight, an explicit similarity solution of this problem is known. The velocity is uniform in region I upstream of the shock, which is a straight line. The velocity suffers an abrupt transition at the shock; it is reduced in magnitude and its direction is rotated. The flow is again uniform downstream of the shock in region II, but parallel to the surface of the wedge. Although we offer a brief derivation of this flow in what follows, we refer to Courant-Friedrichs [1] for a more complete discussion.

The main issue of this note is to investigate the flow when the surface of the wedge is slightly curved. We show that in this case a flow with the same qualittative features still obtains, at least in a bounded region downstream of the vertex. The flow is smooth away from a single shock, which is itself a smooth curve. Of course if a stability result of this type did not hold, one would not trust the mathematical formulation of the problem, as this kind of flow is so readily observable in experiments .

The gas dynamical equations which govern this flow are

$$(1) \qquad \begin{cases} \operatorname{div} (\rho(q) \vec{q}) = 0 \\ \operatorname{curl} \vec{q} = 0 \ . \end{cases}$$

Here $\vec{q}$ is a two-component velocity vector that depends on two real variables x and y , $q = |\vec{q}|$, and $\rho(q)$ is the density expressed as a function of q by Bernoulli's law. For an ideal gas

$$(2) \qquad \rho(q) = \rho_0 \left(1 - q^2 / \hat{q}^2 \right)^{\gamma}$$

where ρ_0 , γ , and $\hat{q}$ are constants; $\hat{q}$ is called the limiting speed as steady state flows exceeding this speed are not possible. For purposes of this exposition we consider only the density relation (2). (A more exact treatment of this problem would involve a 3×3 system in which entropy was the third unknown. Although the 3×3 system is considered throughout [8], we restrict our attention here to (1) for simplicity.)

On performing the differentiations in (1) one obtains a quasi-linear 2×2 system

$$(3) \qquad B_1(q) \frac{\partial q}{\partial x} + B_2(q) \frac{\partial q}{\partial y} = 0$$

where

$$B_1(q) = \begin{pmatrix} c^2(q) - q_1^2 & -q_1 q_2 \\ 0 & 1 \end{pmatrix}$$

and

$$B_2(q) = \begin{pmatrix} -q_1 q_2 & c^2(q) - q_2^2 \\ -1 & 0 \end{pmatrix}$$

In these formulas $c(q)$ is the speed of sound in the gas, given by the explicit formula

$$c^2(q) = -\, q\rho \Big/ \frac{\partial\rho}{\partial q} = \text{const}\left(1 - q^2 \big/ \hat{q}^2\right) . \tag{4}$$

(One must consider the full time dependent equations to see that $c(q)$ is the speed of sound.) Note that the graph of $c(q)$ has the shape indicated in Figure 2.

Much insight can be gained from consideration of the characteristics of (3). Of course (3) is a nonlinear equation, and whether a particular curve is characteristic depends on the Cauchy data. It is obvious from (1) that this system possesses rotational and translational symmetry. Thus it suffices to pose the question of whether the line $\{x = 0\}$ is non-characteristic for (3) at the origin, with respect to some Cauchy data. This will be the case if and only if one may solve explicitly for $\partial q/\partial x$ in (3); i.e., if and only if

$$\det B_1(q) = c^2(q) - q_1^2 \neq 0 \ .$$

On observing that q_1 is the component of $\vec{q}$ normal to the line $\{x=0\}$ we see that a general curve is non-characteristic if and only if

$$c^2(q) - q_N^2 \neq 0 \quad . \tag{5}$$

Two cases should be distinguished here, according to

$$\text{(i)} \quad q < c(q) \qquad \text{or} \qquad \text{(ii)} \quad q > c(q) \quad . \tag{6}$$

If the first option holds in (6), then no matter what the orientation of a curve, we have

$$|q_N| \le q < c(q) \quad ,$$

so that no directions are characteristic. In other words, (3) is an elliptic system, and the flow is called subsonic. On the other hand, if $q > c(q)$, then there are precisely two directions such that

$$q_N = q \cos \theta = c(q) \quad ,$$

where θ is the angle between $\vec{q}$ and the normal. In this case (3) is hyperbolic, and the flow is called supersonic. We consider only supersonic flow in this note. It is clear from Figure 2 that there is a constant q^* such that option (i) or (ii) holds in (6) according to whether $q < q^*$ or $q > q^*$. The constant q^* is called the critical speed.

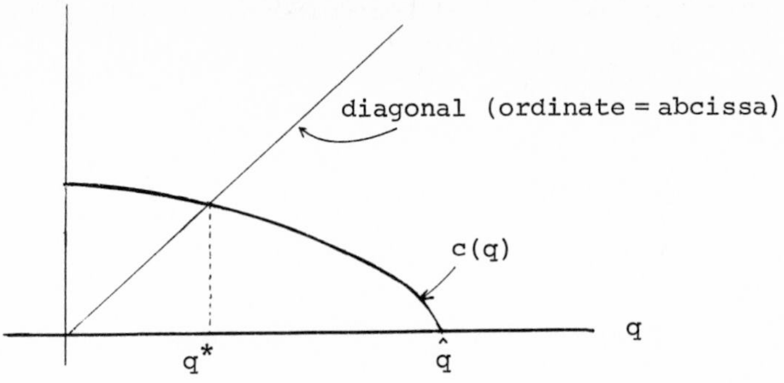

Figure 2

It is obvious that any constant function is a solution of (3). A more interesting special solution, however, is given by two distinct constant states separated by a straight line (see Figure 3). Of course, such a function is discontinuous, so to speak of a solution of (3) one must pass to the weak form of the equations,

$$(7) \qquad \begin{cases} \rho(q)\, q_N = \rho(q')\, q_N' \\ q_T = q_T' \end{cases},$$

where $\vec{q}$ and $\vec{q}\,'$ are the constant states upstream and downstream of the shock, respectively, and q_N, q_T represent the resolution of $\vec{q}$ into normal and tangential components. The second equation here is transparent, but some discussion of the first which states that the flux of gas across the shock is conserved, is appropriate. In Figure 4 we have plotted the flux as a function of q_N, assuming q_T fixed. A simple calculation with (2) shows that the flux across

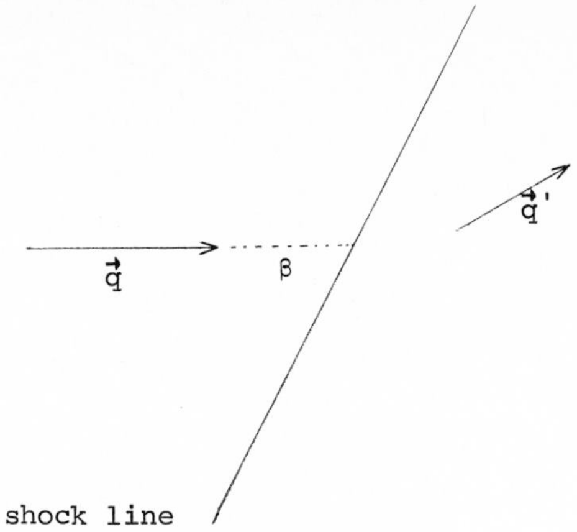

Figure 3

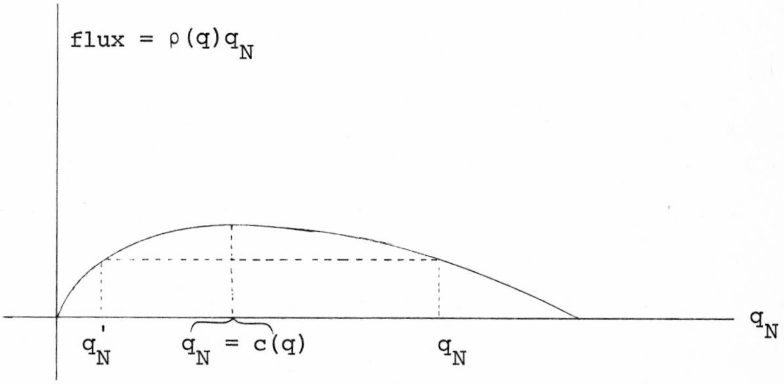

Figure 4

a line attains its maximum when $q_N = c(q)$; i.e., when the line is characteristic. The flux decreases monotonely on either side of this maximum. Thus, given a value for q_N such as shown in the figure, there is a unique value of q_N' (also shown) satisfying (7). The two states thus constructed will yield a weak solution of (3). Let us remark that this solution is only physical if $q_N > c(q)$, as indicated. It follows

(8) $$q_N' < c(q') ,$$

an inequality that will be needed below.

These weak solutions may be used to construct the flow past a straight wedge. Let us consider only the flow past the top half of the wedge, as this is entirely independent of the flow past the bottom edge. To obtain a solution one need only vary the angle β of Figure 3 so that the downstream velocity $\vec{q}$ is parallel to the surface of the wedge. This is possible if the opening angle of the wedge is not too great. In fact, in this case there are two different values of β which make $\vec{q}'$ parallel to the surface of the wedge, one with $\vec{q}'$ subsonic and one supersonic, and both are stable. One is left with a somewhat embarrassing choice between non-unique solutions. It appears that supersonic flow downstream is the more relevant solution physically, though we do not know of a convincing argument for this. In any case, the methods of this note seem to work only for the supersonic case, and we limit our attention to this case.

We now pose the free boundary problem that is the main subject of these notes. (It may be helpful to refer to Figure 5.) We work in the neighborhood of a solution with a straight shock and supersonic flow downstream as described above, which we call the unperturbed problem. Choose a coordinate system as indicated such that the y-axis coincides with the unperturbed shock. Let $x = g(y)$ be the equation of the wedge and let $x = h(y)$ be the (unknown) equation of the shock, the free boundary in the problem. We are looking for a solution of (1) in the region between these two curves, subject to boundary conditions at both sides. Along the wedge we require that the flow be tangential. Along the shock we impose the jump conditions (7). Note that the jump conditions vary from point to point only through changes in the tangent $h'(x)$ to the shock curve, which change the decomposition into normal and tangential components. In other words, along the shock we have Cauchy data $\vec{q} = \vec{Q}(h')$, where Q is some vector valued function of a real variable.

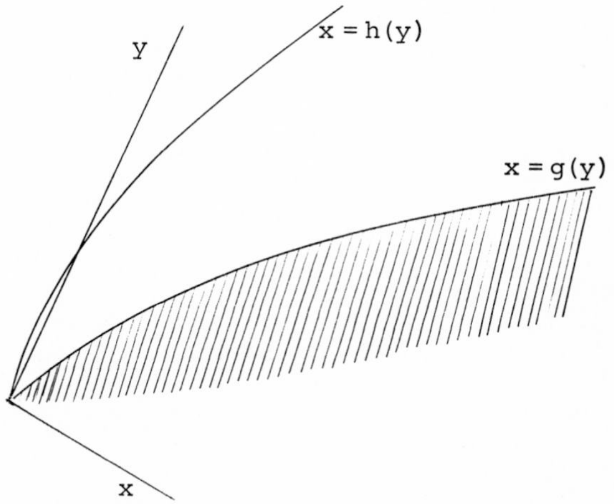

Figure 5

In summary, we have the following problem:

Given $g(y)$, find $h(y)$ and $\vec{q}(x,y)$ in the region

$$\{(x,y) : y > 0 \ , \quad h(y) \le x \le g(y)\}$$

satisfying

$$B_1(q)\,\frac{\partial q}{\partial x} + B_2(q)\,\frac{\partial q}{\partial x} = 0 \tag{9}$$

subject to

$$\text{(10)} \quad \begin{cases} \text{(i)} \quad \vec{q} = \vec{Q}(h') & \text{along } \ x = h(y) \\ \text{(ii)} \quad \langle N_g \, , \vec{q}\rangle = 0 & \text{along } \ x = g(y) \ . \end{cases}$$

Note that the first relation in (10) is a vector equation, so is equivalent to two scalar boundary conditions; the second is a scalar relation.

In stating our main theorem we use the notation $g_0(y) = (\tan\gamma)\,y$ for the equation of the unperturbed wedge. As $x=0$ is the equation of the unperturbed shock, no special notation is required for this.

<u>Theorem</u>: <u>Let</u> $\varepsilon > 0$, $R > 0$ <u>be given. If</u> $g - g_0$ <u>is sufficiently small in the Hölder norm</u> $C^{2+\varepsilon}$, <u>the above problem is soluble on an open neighborhood of the origin containing the disk of radius</u> R . <u>The shock curve</u> $\{x = h(y)\}$ <u>is of class</u> $C^{2+\varepsilon}$ <u>and close to the unperturbed shock.</u>

A simple example shows that an estimate on the $C^{2-\varepsilon}$ norm of $g - g_0$ would not be sufficient. Let a and δ be small, positive parameters and consider

$$g(y) = g_0(y) - \delta(y-a)_+^{2-\varepsilon} \quad , \tag{11}$$

where x_+ equals x or 0 according to the sign of x . The curvature of (11) is unbounded near $y = a$ and of a sign that produces a compression wave. It follows that a continuous solution of (1) near this point is impossible; there must be a shock which terminates on the surface of the wedge at $y = a$, though it may be a very weak shock. For the same reason one can only hope for a solution of this problem in a bounded region downstream of the origin in general. It is shown in [1] that concave portions of the wedge (i.e., regions where $g'' < 0$) tend to produce secondary shocks downstream. We conjecture that only one shock develops downstream of a purely convex wedge, but are at a loss to prove this.

The method of proof for the theorem is based on the following observation: If the shock curve $h(y)$ were known and if one ignores the second boundary condition (10,ii), the remaining problem (9), (10,i) is just a Cauchy problem with data prescribed along the curve $\{x = h(y)\}$. Since we are working close to a flow that is supersonic downstream, the governing equations will be hyperbolic here. Moreover, it follows from (8) that the unperturbed shock is non-characteristic with respect to the constant downstream data of the unperturbed solution. A simple continuity argument shows that the Cauchy problem (9), (10,i) may be solved in a bounded neighborhood of the origin, if h is sufficiently small in C^2. As is shown in [2], the Cauchy data need only be small in C^1, but this data is computed from h'.) Our proof is based on an iteration which assumes $h(y)$ given, solves the Cauchy problem, estimates the size of $\langle N_g, q\rangle$ along the wedge, and recomputes a better approximation for $h(y)$ to use in the next iteration. In point of fact, this iteration is only implicit in our proof, being concealed in the proof of the Nash-Moser theorem to which we appeal.

In solving this Cauchy problem, a certain amount of care is required to assure that the problem is posed in a region that is contained in the domain of dependance of the initial data. Because (9) is non-linear, the domain of dependence will depend on the solution. We refer to Figure 6. In this figure, the line containing OP is the surface of the unperturbed wedge and the line containing OBA_2A_1 the unperturbed shock, or y-axis. A_1P and A_2P are the two characteristics of the unperturbed solution of (3) which terminate at P, and OB is the projection of OP onto the y-axis. Let I_1 be an interval of the y-axis which contains OA_1

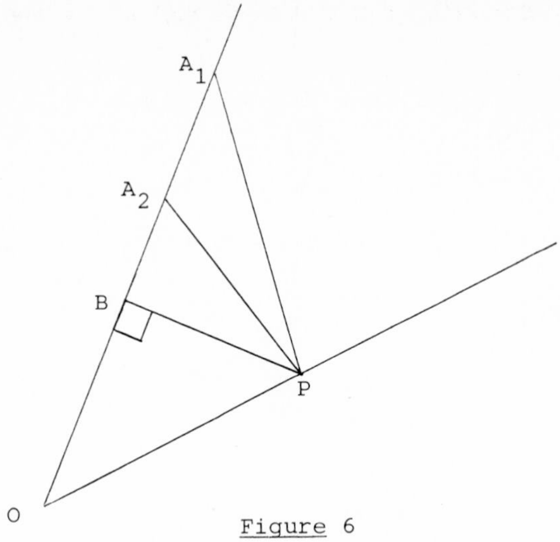

<u>Figure</u> 6

in its interior, and let I_2 be OB . Provided $\|h : C^2\|$ is sufficiently small, the interval OP will be contained in the interior of the domain of dependance of the curve

$$(h(y), y) : y \in I_2$$

for the solution to the Cauchy problem (9), (10,i). Moreover, by making $g - g_0$ small we can arrange that the perturbed surface of the wedge

$$\{(g(y), y) : y \in I_2\}$$

is also contained in this domain of dependance.

Having made the above construction we may define a non-linear mapping $\Phi : \mathcal{O} \to C^1(I_2)$, where

$$\mathcal{O} = \{(h , g) \in C^2(I_1) \times C^2(I_2) : \|h\| < \varepsilon , \|g\| < \varepsilon\} ,$$

as follows. Here ε is a constant, sufficiently small to ensure that the Cauchy problem is soluble, and consistent with the above construction. Given $(h , g) \in \mathcal{O}$, let $\vec{q}$ be the solution of the Cauchy problem (9), (10,i) and let

(12) $\Phi(h, g) = \langle Ng, \vec{q}\rangle \mid \{x = g(y)\}$.

More properly (12), a function of y , should be written out

$$q_1(g(y), y) - g'(y)q_2(g(y), y) \quad .$$

We define this operator because solving the free boundary (9), (10) is equivalent to solving

(13) $\Phi(h, g) = 0$.

In [8] we solve (13) by an appeal to the Nash-Moser implicit function theorem. Two kinds of hypotheses must be verified to apply this theorem. First one must estimate very carefully the norm of Φ and its differentials $d\Phi$, $d^2\Phi$ as operators from C^{ℓ} to $C^{\ell - m}$, where m is a fixed parameter and ℓ varies over an interval. For the problem at hand this is tedious but entirely classical, using only energy estimates. Second, one must show that $d\Phi$ is invertible and obtain estimates for the inverse similar to those needed for $d\Phi$ itself. A little care is required to invert $d\Phi$ uniformly on a range of C^{ℓ} spaces, which is done in §§7 - 10 of [8]. This is the least obvious part of the proof and we have little in the way of motivating remarks. The operator Φ is defined on C^2 , but because the Hölder spaces with integer exponents are badly behaved, we are only able to solve (13) if we have an estimate for $\|g - g_0 : C^{2+\varepsilon}\|$. Indeed, even to get this result we must appeal to the recent strengthened version of the Nash-Moser theorem due to Hörmander [3] that was unavailable when [8] was written. However, the methods needed to verify the hypotheses are in no way different.

Rather than enter into the details of the proofs of [8], it is perhaps more useful to clarify the loss of derivatives phenomenon in this problem which precludes an application of the ordinary implicit function theorem in C^ℓ . This is better done with a simplified example rather than the original problem. Consider a mapping Φ given by $\varphi \to u(1, \cdot)$ where u is the solution of a quasi-linear Cauchy problem

$$(14) \qquad \begin{cases} \dfrac{\partial u}{\partial x} + A(u)\dfrac{\partial u}{\partial y} = 0 \\ \\ u(0, y) = \varphi(y) \quad . \end{cases}$$

Suppose that φ is restricted to a small neighborhood of zero in order to guarantee that (14) is soluble up to $x = 1$. Φ maps C^ℓ into itself for any ℓ , assuming that A is smooth with respect to its argument. The differential $d\Phi(\psi)$ may be obtained by solving the linear Cauchy problem

$$(15) \qquad \begin{cases} \dfrac{\partial v}{\partial x} + A(u)\dfrac{\partial v}{\partial y} + \left(A'(u)\dfrac{\partial u}{\partial y}\right) v = 0 \\ \\ v(0, y) = \psi(y) \quad . \end{cases}$$

We have $d\Phi(\psi) = v(1, \cdot)$. The best estimate available for v , and hence for $d\Phi(\psi)$, is

$$\|v\|_\ell \le C \left\{\|\psi\|_\ell + \|\varphi\|_{\ell+1} \|\psi\|_0\right\}$$

One cannot obtain an estimate with fewer derivatives on φ , because to have $v \in C^\ell$ we must have the coefficients in (15) belonging to C^ℓ , which requires that $u \in C^{\ell+1}$, which in turn implies that

$\varphi \in C^{\ell+1}$. Thus, although $d\Phi$ is a bounded operator on C^{ℓ} provided $\varphi \in C^{\ell+1}$, the norm of this operator depends on $\|\varphi : C^{\ell+1}\|$. The inverse $d\Phi^{-1}$ exhibits similar behavior, and the second differential is slightly worse - to estimate the C^{ℓ} norm of $d^2\Phi(\psi_1, \psi_2)$ we need $\|\psi_1\|_{\ell+1}\|\psi_2\|_0$ and $\|\psi_1\|_0\|\psi_2\|_{\ell+1}$ as well as the terms in $\|\varphi\|_{\ell+1}$ that were to be expected. That is, even with $\varphi \in C^{\ell+1}$ the second differential is not a bounded bilinear mapping on C^{ℓ} . This is the mechanism by which derivations are lost in the problem of this note - mild compared to some applications of the Nash-Moser theorem but apparently sufficient to require the hard implicit function theorem.

REFERENCES

[1] R. Courant and K.O. Friedrichs, Supersonic flow and shock waves. Interscience, New York, 1948.

[2] A. Douglas, "Some existence theorems for hyperbolic systems of partial differential equations in two independent variables", Comm. Pure Appl. Math. 5 (1952), pp. 119-154.

[3] L. Hörmander, "The boundary problems of physical geodesy," Archive Rat. Mech. Anal. 62 (1976), pp. 1-52.

[4] J. Möser, "A new technique for the construction of solutions of non-linear differential equations", Proc. Nat. Acad. Sci. 47 (1961), pp. 1824-31.

[5] J. Nash, "The embedding problem for Riemannian manifolds", Annals of Math. 63 (1956), pp. 20-63.

[6] D. Schaeffer, "The capacitor problem", Indiana Math. J. 24 (1975), pp. 1143-1167.

[7] D. Schaeffer, "A stability theorem for the obstacle problem", Advances in Math. 17 (1973), pp. 34-47.

[8] D. Schaeffer, "Supersonic flow past a nearly straight wedge", Duke Math. J. 43 (1976), pp. 637-670.

SINGULARITIES IN NON-LINEAR WAVES OF KLEIN-GORDON TYPE

Michael C. Reed

Department of Mathematics
Duke University

The main purpose of these lectures is to outline a proof that the propagation of singularities for the equation

$$(1) \qquad \begin{aligned} u_{tt} - u_{xx} &= f(x,u,u_x,u_t) \qquad x \in \mathbb{R} \\ u(x,0) &= f_1(x) \\ u_t(x,0) &= f_2(x) \end{aligned}$$

is the same as for the corresponding linear equation $u_{tt} - u_{xx} = 0$. That is, singularities in the initial data can propagate only along light rays. Details will appear in [6] . Of course, we must decide what we mean by "singularity" since we cannot hope to solve (1) , even locally, for arbitrary distributional initial data. For the moment, we will say that u is "smooth" at $\langle x,t\rangle$ if it is C^∞ there and say that u is "singular" if it is not C^∞ . We will assume that f is C^∞ ; otherwise f itself would produce new singularities. And, we will make the assumption that

$$f(x,0,0,0) = 0 \quad \text{for all } x \quad .$$

This assumption insures that data with compact support continue to have compact support in the future. This is convenient for the proof, but it is not related to the question of propagation of singularities and could be avoided with a little more work. Otherwise, we need no structural assumptions on f at all. The result says that $u(x,t)$ will be C^∞ everywhere except at points $\langle x,t \rangle$ where one of the backward light rays through $\langle x,t \rangle$ intersects a singular point of the initial data. This is true locally (in time) and it is true globally if the solution is global. Of course, (1) will have a global solution only under special additional hypotheses on f.

Before beginning let me say a few words about why one should be interested in this problem at all. One of the most interesting physical phenomena of non-linear wave motion is the creation and absorbtion of singularities. If two smooth water waves meet, a line of cusp singularities can form which propagates for a short time and then is absorbed so that the surface of the water is again smooth. Of course, the breaking of waves is an even more spectacular example of the production of new singularities, but I like the cusp example better because in that case it is clear that a statistical model will not suffice. Well, it is the job of the mathematician, applied mathematician, mathematical physicist, to find models for such phenomena, and as you know, it is not an easy task. Partly we are handicapped by dependence on some of the highly efficient tools developed to solve linear problems. For example, one typically solves linear hyperbolic problems by identifying a self-adjoint infinitesimal generator A and then propagating the initial data with the corresponding group $U(t) = e^{-itA}$. If the data has a certain degree of smoothness, which usually means it is the domain of A^N for some N, then the solution will have the same degree of smoothness since $U(t)$ takes the domain of A^N into itself. Secondly, because of the non-linearity, it is not clear how to use the powerful tools of localization and then the Fourier transform. To see how hard these problems really are, I pose the following problem for you: Find a hyperbolic equation in two space dimensions which can produce and then absorb cusp singularities.

My own interest in these problems was stimulated by the fact that certain hyperbolic conservation laws can force the production of new singularities even if the data are smooth. (See, for example Lax and Glimm [4]). I asked myself what is the behavior of singularities for the Klein-Gordon type equations which occur so frequently in mathematical physics. At least in one space dimension, the answer is "just the same as in the linear case". The problem is much harder in two and three space dimensions and remains unsolved.

I. GLOBAL EXISTENCE

In order to solve (1), we first reformulate it as a first order system: let $v = \frac{1}{2}(u_x + u_t)$ and $w = \frac{1}{2}(u_x - u_t)$. Then, denoting the indefinite integral by

$$[I(u)](x) = \int_{-\infty}^{x} h(y)\,dy$$

(1) becomes

$$\begin{aligned} v_t - v_x &= \tfrac{1}{2} g_{v,w} \\ w_t + w_x &= -\tfrac{1}{2} g_{v,w} \\ v(x,0) &= v_0(x) \\ w(x,0) &= w_0(x) \end{aligned} \qquad (2)$$

where $v_0(x) = \frac{1}{2}(f_1' + f_2)$, $w_0(x) = \frac{1}{2}(f_1' - f_2)$ and

$$g_{v,w}(x,t) = f(x, [I(v + w)](x,t), v(x,t) + w(x,t), v(x,t) - w(x,t)).$$

And, in order to study (2) we study the corresponding integral equations:

$$\begin{aligned} v(x,t) &= v_0(x + t) + \tfrac{1}{2}\int_0^t g_{v,w}(x + (t-s), s)\,ds \\ w(x,t) &= w_0(x - t) - \tfrac{1}{2}\int_0^t g_{v,w}(x - (t-s), s)\,ds \end{aligned} \qquad (3)$$

Notice that since we are interested in initial data for (1) which have compact support, we not only require the data, v_0, w_0 , to have compact support but also to satisfy

$$\int_{-\infty}^{\infty} v_0(y) + w_0(y)\,dy = 0 \qquad .$$

The solution of (3) should then satisfy

$$\text{(S)} \qquad \int_{-\infty}^{\infty} v(y,t) + w(y,t)\,dy = 0 \qquad \text{for all } t$$

so that

$$u(x,t) = \int_{-\infty}^{\infty} v(y,t) + w(y,t)\,dy$$

has the right support properties. For lack of a better name, I will call this "property S" . It will be crucial later.

Now, the usual contraction mapping argument shows that if the initial data $\langle v_0, w_0 \rangle$ is in $C_0^N \times C_0^N$, $N = 0,1,2,\ldots,$ and satisfies property S , then there is a local solution $\langle v(x,t), w(x,t) \rangle$ of (3) which is in $C_0^N \times C_0^N$ and satisfies property S for each t for which it exists. There is a slight abuse of notation here which I should clarify. $v_0(x) \in C_0^N$ means that v_0 is N times continuously differentiable with compact support. $v(x,t) \in C_0^N$ means that v is C^N in both x and t (for some time interval) and that v has compact support in x for each fixed t .

As you know I'm sure, the question of global existence in t is more delicate and depends on special properties of f . There are various kinds of hypotheses on f and the initial data which guarantee global existence: (1) the existence of a positive conserved energy [1], [7]; (2) weak non-linearities [2],[5]; (3) high degree of non-linearity and small initial data [8],[9]; (4) special conservation laws and symmetry properties of the initial data [3] . Information about these methods can be found in the references. There is just one point which I wish to make. All cases that I know of have the following property.

If one can prove global existence in $C_0 \times C_0$ then one automatically gets an estimate on the size of the solution in terms of the size of the initial data:

$$(4) \qquad |v(x,t)| + |w(x,t)| \le d(t)\{\sup_x |v_0(x)| + \sup_x |w_0(x)|\}$$

where $d(t)$ may go to ∞ as $t \to \pm\infty$. The point is that from this it automatically follows that if the data is in $C_0^N \times C_0^N$ then the solution is global in $C_0^N \times C_0^N$. Let me explain why this is. Suppose the data is in $C_0^N \times C_0^N$. Then, by the contraction mapping argument, the solution is in $C_0^N \times C_0^N$ for short times and by our assumption exists globally in $C_0 \times C_0$. To show it is global in $C_0^N \times C_0^N$ we must show that the first N derivatives, which exist locally, do not blow up in finite time. To see this differentiate the equations (3) , for example with respect to x .

$$(5) \qquad \begin{aligned} v_x &= v_0'(x + t) + \tfrac{1}{2}\int_0^t f_1 + (v + w)f_2 + (v_x + w_x)f_3 + (v_x - w_x)f_4 ds \\ w_x &= w_0'(x - t) - \tfrac{1}{2}\int_0^t f_1 + (v + w)f_2 + (v_x + w_x)f_3 + (v_x - w_x)f_4 ds \end{aligned}$$

where f_i denotes the partial with respect to the ith variable. On any finite time interval everything on the right hand side of (5) is a priori bounded except for v_x and w_x. By iteration (Gronwall's inequality), v_x and w_x are a priori bounded too. A similar proof works for v_t , w_t , and for higher derivatives. The point is that at each stage the highest order terms occur linearly and the lower order terms have already been a priori bounded. Thus, if (3) has global solutions in $C_0 \times C_0$ satisfying (4), then it has global solutions in $C_0^N \times C_0^N$ for each N .

There is one other fact about (3) which is very important because it expresses the domain of dependence of the initial data through an a priori estimate. By iterating the integral equations (3) and the integral equations obtained by differenting some number of times with respect to x or t, one can obtain the estimate

$$(6) \qquad |D_x^N v(x,t)| \le d(t,K)\left\{\sum_{j=0}^{N} \sup_{y\in\mathbb{R}} |D_y^j v_0(y)| + \sup_{y\in\mathbb{R}} |D_y^j w_0(y)|\right\}$$

where d is a constant depending on time and the support of the initial data, K. This estimate is unsatisfactory because the sups on the right are over all of $\mathbb{R}$ rather than just over $y \in [x-t,\ x+t]$. This happens in the iteration because of the non-local term $I(v + w)$ in $g_{v,w}$. One can improve (6) by the following two-step process. First one shows that if $\langle v_0^*, w_0^*\rangle$ is another pair of initial data such that

$$(7) \qquad v_0^*(y) = v_0(y),\ w_0^*(y) = w_0(y),\ I(v_0^* + w_0^*)(y) = I(v_0 + w_0)(y)$$

for $y \in [x-t,\ x+t]$, then the solutions $\langle v,w\rangle$ and $\langle v^*,w^*\rangle$ are equal in the triangle shown in Figure 1.

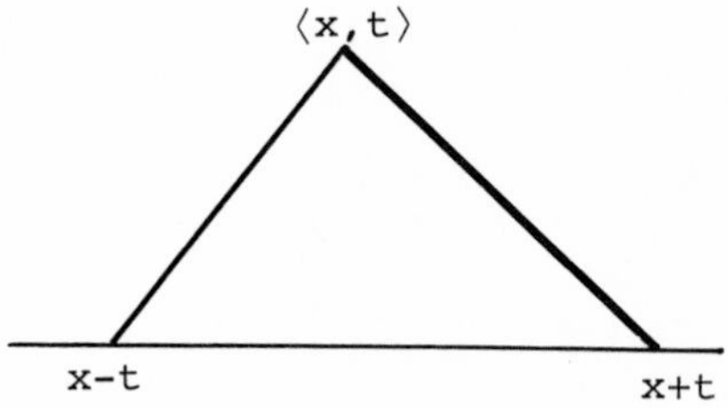

Figure 1

Second, given $\langle v_0,w_0\rangle$ on $[x-t,\ x+t]$ one shows that $\langle v_0,w_0\rangle$ can be extended to functions $\langle v_0^*,w_0^*\rangle$ of compact support on all of $\mathbb{R}$ in such a way that (7) holds and

(8) $$\sup_{y\in\mathbb{R}}|D^k v_0^*(y)| \le F_k\Big(\sup_{x-t\le y\le x+t}|v_0(y)|,\ldots,\sup_{x-t\le y\le x+t}|D_Y^N v_0(y)|\Big)$$

for $k = 0,1,\ldots,N$, where F_k is a continuous function which depends only on N. Combining (6) and (8) and similar estimates for w_0^*, we have

(9) $$|D_x^N v(x,t)| \le d(t,K)F\Big(\ldots,\sup_{x-t\le y\le x+t}|D_Y^j v_0(y)|,\ldots,\sup_{x-t\le y\le x+t}|D_Y^j w_0(y)|,\ldots\Big)$$

for some fixed continuous F.

II. PROPAGATION OF AN INTERVAL OF SINGULARITIES

Let us start by considering the following special case. Suppose that v_0 and w_0 have support in $[-b,b]$, are continuous in $[-a,a]$, and C^∞ outside of $[-a,a]$. Let $\mathfrak{m}$ be the mapping on pairs of functions given by $\mathfrak{m}: \langle v,w\rangle \to \langle M_1, M_2\rangle$ where

(10)
$$M_1(x,t) = v_0(x+t) + \tfrac{1}{2}\int_0^t g_{v,w}(x+(t-s),s)\,ds$$

$$M_2(x,t) = w_0(x-t) - \tfrac{1}{2}\int_0^t g_{v,w}(x-(t-s),s)\,ds$$

We must choose a space on which $\mathfrak{m}$ is a contraction and we want the pairs of functions in the space to have the smoothness properties that we expect so that the fixed point of $\mathfrak{m}$ (the solution of (3)) will have the right smoothness. Thus, before we begin, we must decide what the smoothness properties are that we are trying to prove.

Look at Figure 2 . If we expect the same behavior as in the linear case, the solution should be zero in region V and C^∞ in regions IVa, IVb, and IVc . But, what about the regions I, II, III, where we expect to prove that the singularities are propagating? The

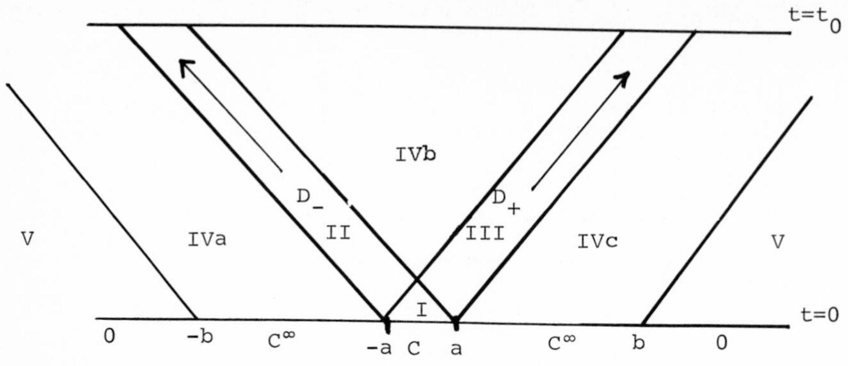

Figure 2

first guess is that the solution $\langle v,w \rangle$ will just be continuous in I, II, and III . But this guess is wrong! The correct smoothness properties are as follows. v and w are only continuous in region I . Let D_+ and D_- denote derivation in the directions indicated in Figure 2 . Then v and w will be C^∞ in the D_- direction in region II but only continuous in all other directions. And, in region III , v and w will be C^∞ in the D_+ direction and only continuous in all other directions. These rather funny properties are in fact just the properties of the corresponding linear equation ($g_{v,w} = 0$) since $v_0(x + t)$ is constant (and therefore C^∞) in the D_- direction and $w_0(x - t)$ is constant (and therefore C^∞) in the D_+ direction. As we will see below, if we only assumed continuity in regions II and III, then $\mathfrak{m}$ would not be a contraction and the proof would not go through.

So, let $Y(N,t_0)$, $N \geq 1$, denote the set of functions h with the following properties:

(a) $h(x,t)$ is defined for $t \in [0,t_0]$ and $x \in \mathbb{R}$, is continuous, and is zero in region V .

(b) $h(x,t)$ is N times continuously D_+ and D_- differentiable in regions IVa, IVb, IVc .

(c) $h(x,t)$ is N times continuously D_- differentiable in region II

(d) $h(x,t)$ is N times continuously D_+ differentiable in region III .

Let $\|\cdot\|^{(N)}$ be the natural norm on $Y(N,t_0)$ so that it is a Banach space. Let $X(N,t_0)$ denote the set of pairs $\langle v,w\rangle \in Y(N,t_0) \times Y(N,t_0)$ such that $\langle v,w\rangle$ satisfies property S , $v(x,0) = v_0(x)$, $w(x,0) = w_0(x)$, $\|v - v_0(x,t)\|^{(N)} \le 1$, $\|w - w_0(x,t)\|^{(N)} \le 1$.

We must show that $\mathfrak{m}$ is a contraction on $X(N, t_0)$ for t_0 small enough. The proof goes exactly as expected. There is just one crucial point to check: namely, that $\mathfrak{m}$ takes X into itself as far as the differentiability properties are concerned. That is, we must check that if v and w are in $Y(N, t_0)$, then so are M_1 and M_2 . We do this in three steps. First, define $C_+(x,t)$ and $C_-(x,t)$ to be the straight lines from $\langle 0, x-t\rangle$ to $\langle x,t\rangle$ and $\langle 0, x+t\rangle$ to $\langle x,t\rangle$ respectively. Then, we can rewrite (10) as

$$M_1(x,t) = v_0(x + t) + \frac{1}{2\sqrt{2}} \int_{C_-(x,t)} g_{v,w}(\xi)\,d\xi$$

$$M_2(x,t) = w_0(x - t) - \frac{1}{2\sqrt{2}} \int_{C_+(x,t)} g_{v,w}(\xi)\,d\xi$$

LEMMA 1. <u>Suppose that</u> h <u>is in</u> $Y(N, t_0)$ <u>and define</u>

$$k_\pm(x,t) = \int_{C_\pm(x,t)} h(\xi)\,d\xi$$

<u>Then</u> $k_\pm$ <u>are in</u> $Y(N, t_0)$.

Let me indicate why this lemma is true by considering k_- at a point $\langle x,t\rangle$ in region IVb . It is clear that the D_- derivative exists and

$$D_-k_- = h(x,t) \quad .$$

To compute D_+k_- we first rewrite the difference quotient:

$$\frac{1}{\varepsilon}\left\{k_-(x+\varepsilon/\sqrt{2},\ t+\varepsilon/\sqrt{2}) - k_-(x,t)\right\}$$

$$= \frac{1}{\varepsilon}\left\{ \int_{C_-(x+\varepsilon/\sqrt{2},\ x+t/\sqrt{2})} h(\xi)d\xi - \int_{C_-(x,t)} h(\xi)d\xi \right\}$$

$$= \int_{C_-(x,t)} \frac{h(\xi_1+\varepsilon/\sqrt{2},\ \xi_2+\varepsilon/\sqrt{2})-h(\xi_1,\xi_2)}{\varepsilon}$$

$$+ \frac{1}{\varepsilon}\int_{C_-(x+t+\varepsilon/\sqrt{2},\ \varepsilon/\sqrt{2})} h(\xi)d\xi$$

See Figure 3. Now, we can take the limit of the difference quotient in the first integral since for all ξ on $C_-(x,t)$, h is D_+ continuously differentiable. And, the limit of the second term is $h(x+t,\ 0)$. Thus k_- is D_+ differentiable at $\langle x,t\rangle$ and:

$$D_+k_- = \int_{C_-(x,t)} D_+h(\xi)d\xi + h(x+t,\ 0) \quad .$$

Thus we see that k_- is both D_- and D_+ differentiable at $\langle x,t\rangle$. And, we can now continue in this manner taking higher derivatives since h is D_+ differentiable at $\langle x+t,\ 0\rangle$ because $x+t > a$ (since $\langle x,t\rangle$ is in IVb).

A similar proof works for k_+ in region IVb and for k_+ and k_- in other regions. The point to notice is that we could never have proven that k_- is D_+ differentiable at $\langle x,t\rangle$ if we hadn't known that h is D_+ is differentiable all along $C_-(x,t)$, in particular in region III.

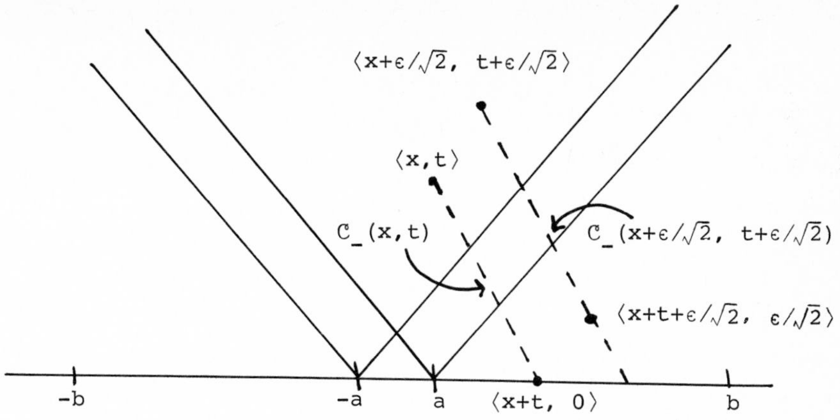

Figure 3

LEMMA 2. *Suppose that* v *and* w *are in* $Y(N,t_0)$. *Then* $I(v+w)$ *is in* $Y(N,t_0)$.

Let us again consider a point $\langle x,t\rangle$ in region IVb , see Figure 4.

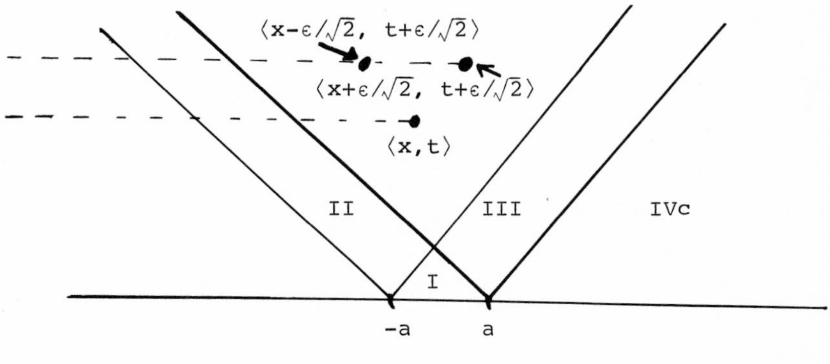

Figure 4

$$\frac{1}{\varepsilon}\left\{I(v+w)(x-\varepsilon/\sqrt{2},\ t+\varepsilon/\sqrt{2}) - I(v+w)(x,t)\right\} =$$

$$\int_{-\infty}^{x} \frac{1}{\varepsilon}\left\{v(y-\varepsilon/\sqrt{2},\ t+\varepsilon/\sqrt{2})-v(y,t)\right\} + \frac{1}{\varepsilon}\left\{w(y-\varepsilon/\sqrt{2},\ t+\varepsilon/\sqrt{2})-w(y,t)\right\}dy$$

So taking limits we get:

$$D_-I(v+w)(x,t) = \int_{-\infty}^{x} D_-v(y,t) + D_-w(y,t)\,dy$$

since v and w are D_- differentiable in region II. To show that $I(v+w)$ is D_+ differentiable at $\langle x,t\rangle$ we write:

$$\frac{1}{\varepsilon}\left\{I(v+w)(x+\varepsilon/\sqrt{2},\ t+\varepsilon/\sqrt{2}) - I(v+w)(x,t)\right\}$$

$$= \int_{-\infty}^{x} \frac{1}{\varepsilon}\left\{v(y-\varepsilon/\sqrt{2},\ t+\varepsilon/\sqrt{2})-v(y,t)\right\} + \frac{1}{\varepsilon}\left\{w(y-\varepsilon/\sqrt{2},\ t+\varepsilon/\sqrt{2})-w(y,t)\right\}dy$$

$$+ \frac{1}{\varepsilon}\int_{x-\varepsilon/\sqrt{2}}^{x+\varepsilon/\sqrt{2}} v(y,t+\varepsilon/\sqrt{2}) + w(y,t+\varepsilon/\sqrt{2})\,dy$$

so taking limits

$$D_+I(v+w)(x,t) = \int_{-\infty}^{x} D_-v(y,t) + D_-w(y,t)\,dy + \sqrt{2}(v(x,t) + w(x,t))$$

since v and w are D_- differentiable in region III.

Notice that this proof that $I(v+w)$ is D_+ differentiable at $\langle x,t\rangle$ would not work if $\langle x,t\rangle$ were in region IVc since then the path of integration would cross region III where v and w are not D_- differentiable. But, we are saved by property S which says that

$$\int_{-\infty}^{x} v(y,t) + w(y,t)\,dy = -\int_{x}^{\infty} v(y,t) + w(y,t)\,dy$$

so in region IVc we can use the same idea, but integrate in from the right instead of the left. The only region where this trick doesn't work is in region 1, but there there is nothing to prove since $I(v+w)$ is only required to be continuous.

Now we can see why $\mathfrak{m}$ preserves the right differentiability properties. If v and w are in $Y(N,t_0)$, then by Lemma 2, $I(v+w)$ is in $Y(N,t_0)$. Thus, since f is C^∞,

$$g_{v,w} = f(x,I(v+w),\ v+w,\ v-w)$$

is in $Y(N,t_0)$ too. But,

$$M_1(x,t) = v_0(x+t) + \frac{1}{2\sqrt{2}}\int_{C_-(x,t)} g_{v,w}(\xi)\,d\xi$$

$$M_2(x,t) = w_0(x-t) - \frac{1}{2\sqrt{2}}\int_{C_+(x,t)} g_{v,w}(\xi)\,d\xi$$

so, by Lemma 1 and the fact that $v_0(x+t)$ and $w_0(x-t)$ are in $Y(N,t_0)$, we conclude that $M_1(x,t)$ and $M_2(x,t)$ are in $Y(N,t_0)$.

This is the idea of the proof that $\mathfrak{m}$ preserves the right differentiability properties. The actual proof that $\mathfrak{m}$ is a contraction is somewhat longer (but not difficult) since one must prove estimates along with Lemmas 1 and 2, verify that $\mathfrak{m}$ preserves property S, etc. And, in the case of global existance, one must use the estimates and iteration arguments to show that the rather complicated norm on $Y(N,t_0) \times Y(N,t_0)$ does not blow up in finite time.

THEOREM 1. Suppose that f is a C^∞ function of all of its variables and $f(x,0,0,0) = 0$. Suppose that $\langle v_0(x), w_0(y)\rangle$ has support in the interval $[-b,b]$, is continuous in $[-a,a]$, is C^N with bounded derivatives in $(-\infty,-a)$ and (a,∞) and satisfies $\int_{-\infty}^{\infty} v_0(y) + w_0(y)\,dy = 0$. Then there is a $t_0 > 0$ so that (3) has a unique solution in $Y(N,t_0) \times Y(N,t_0)$ which satisfies property S for each $0 \le t \le t_0$. If f is such that (3) has global solutions in $C_0 \times C_0$ then $\langle v,w\rangle$ is in $Y(N,t_0) \times Y(N,t_0)$ for all t_0.

III. PROPAGATION OF A COMPACT SET OF SINGULARITIES

The main difficulty in extending the result of the previous section to the general case of a compact set of singularities is the nonlocality of the term $I(v+w)$. Recall that in the sketch of the proof of Lemma 2 we needed to use property S to integrate in from both the right and the left in order to prove that if v and w have the right differentiability properties then $I(v+w)$ does too. This difficulty is avoided by proving that the solution $\langle v,w\rangle$ at a point $\langle x,t\rangle$ depends only on the initial data in the interval $x-t\le y\le x+t$. In fact, one needs slightly more than this. For the iteration arguments (which we have been brushing under the rug all along) we need estimates analogous to (9) . For example, if $\langle x,t\rangle$ is in IVb then we need estimates of the form

$$|D_+v(x,t)| \le \tag{11}$$

$$d(t,K)F\Big(\sup_{y\in[-a,a]}|v_0(y)|,\ \sup_{y\in[-a,a]}|w_0(y)|,\ \sup_{\substack{y\notin[-a,a]\\ y\in[s-t,x+t]}}|D_y v_0(y)|,\ \sup_{\substack{y\notin[-a,a]\\ y\in[x-t,x+t]}}|D_y w_0(y)|\Big)$$

and analogously for higher derivatives. These estimates are proven by the technique sketched at the end of Section I . Namely, first one shows that the solution at $\langle x,t\rangle$ doesn't change if the data are changed outside of $[x-t,\ x+t]$. Then one extends the data $\langle v_0,w_0\rangle$ restricted to $[x-t,\ x+t]$ to data on the whole line which allows one to replace the nonlocal estimate one gets from iteration with a local estimate like (11) .

Now, let us outline the argument in the case of two intervals of singularities, (a_1,a_2) and (b_1,b_2) , (see Figure 5) . The solution will have the right differentiability properties in regions IVa, IIa, IVb, IIIa, IVf, Ia, because the solution at a point $\langle x,t\rangle$ in these regions is unaffected by the presence of the interval of singularities (b_1,b_2) . Similarly, everything is all right in the regions IVf, IIc, IVd, IIIc, IVe, Ic, again because of Theorem 1. That leaves IIb, IVc, IIIb, and Ib . But in IIb, IVc, and IIIb , all points can be reached by integrating from either $x = -\infty$ or $x = +\infty$ only

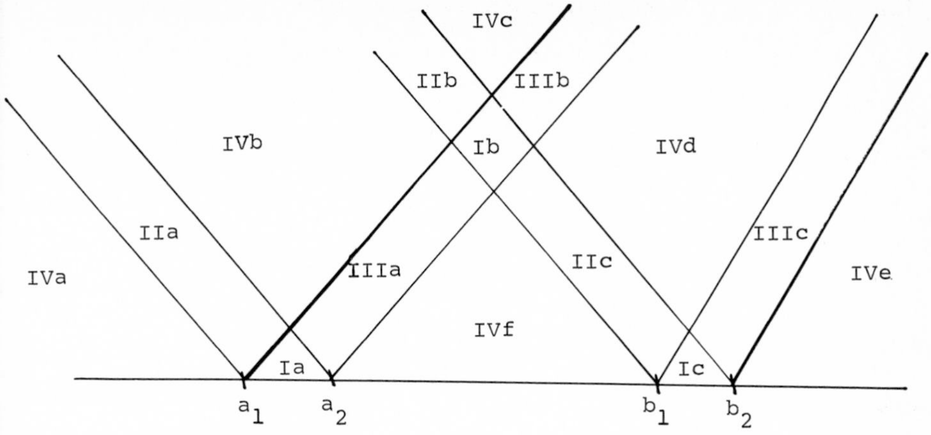

Figure 5

over diagonal strips in one direction or the other, not both. Thus the argument of Lemma 2 goes through in these regions and this plus an iteration argument proves that everything is all right in IIb, IVc, and IIIb. That leaves Ib. But we don't have to prove anything in Ib since the solution is only required to be continuous there.

I will spare you the picture of three intervals of singularities, but if you draw the figure you will immediately see that once you know the theorem for two intervals of singularities, you can get it for three intervals by arguing as above. Continuing in this way one gets the theorem for n intervals of singularities. A simple point set topology argument extends the result to an arbitrary compact set K of singularities: since $\mathbb{R}\backslash K$ is open, we can write $\mathbb{R}\backslash K = \cup O_j$ where the O_n are open intervals. Then one uses the n interval result on $[-b,b] \cap (\mathbb{R}\backslash \overset{n}{\cup} O_j)$ for each n.

There are two final steps in the proof which I should mention. First, since we have the result for each N, we know that, at any point $\langle x,t\rangle$ for which a backward light ray does not intersect k, v and w are infinitely often D_+ and D_- continuously differentiable.

Localizing about $\langle x,t\rangle$ and using the Fourier transform, one easily sees from this that v and w are C^∞ at $\langle x,t\rangle$. Secondly, one must translate the results for v and w into results for u. We want that u be C^∞ in the right places and that u satisfy (1) where it is C^∞. This is not completely trivial because of the non-local definition of $u = I(v+w)$. The final result is:

THEOREM 2. *Let f be a C^∞ function which satisfies* $f(x,0,0,0) = 0$. *Let K be a compact subset of $\mathbb{R}$, f_1 a continuously differentiable function on $\mathbb{R}$ with compact support, f_2 a continuous function with compact support, such that f_1 and f_2 are C^∞ with bounded derivatives on $\mathbb{R}\setminus K$. Then $u = I(v+w)$ is C^∞ at all $\langle x,t\rangle$ such that no forward or backward light ray intersects K and u satisfies* (1). *This is true locally (in t), and if f is such that* (3) *has global solutions in $C \times C$, then it is true globally.*

* * * * *

I am grateful to Indiana University for supporting these lectures and John Chadam for inviting me. I was fortunate to have Dennis Pixton of Indiana University and Robert Kaufman of the University of Illinois in the audience. Each pointed out a mistake in the proof which has been corrected in this manuscript. Chip Berning of Duke University also made many helpful comments.

BIBLIOGRAPHY

[1] Browder, F., "On non-linear wave equations", Math. Zeit. 80 (1962) 249-264.

[2] Chadam, J., "Global solutions of the Cauchy problem for the (classical) coupled Maxwell-Dirac equations in one space dimension" J. Func. Anal. 13 (1973), 173-184.

[3] Chadam, J. and R. Glassey, "On certain global solutions of the Cauchy problem for the (classical) coupled Klein-Gordon-Dirac equations in one and three space dimensions", Arch. Rat. Mech. Anal 54 (1974), 223-237.

[4] J. Glimm and P. Lax, Decay of Solutions of Systems of Nonlinear Hyperbolic Conservation Laws, Amer. Math. Soc. Memoir 101 .

[5] Reed, M., Abstract Non-Linear Wave Equations, Springer Lec. Notes in Math. 507.

[6] __________, "Propogation of singularities for non-linear wave equations in one dimension", Arch. Rat. Mech. Anal. (to appear)

[7] Segal, I., "Non-linear semi-groups", Ann. Math. 78 (1963), 339-364.

[8] __________, "Dispersion for non-linear relativistic equations, II," Ann. Sci. Ecole Norm. Sup. (4) I, (1968), 459-497.

[9] Strauss, W. "Nonlinear Scattering Theory", in Scattering Theory in Mathematical Physics, ed. J. A. Lavita and J. P. Marchand. Reidel, Holland, 1974, 53-78.

BIFURCATIONS OF DYNAMICAL SYSTEMS AND NONLINEAR OSCILLATIONS IN ENGINEERING SYSTEMS

Philip J. Holmes
Institute of Sound and Vibration Research,
Southampton*

and

Jerrold E. Marsden
Department of Mathematics,
University of California, Berkeley
Department of Mathematics,
Heriot-Watt University, Edinburgh

Dedicated to Eberhard Hopf.

INTRODUCTION

This paper analyzes recent qualitative methods for partial differential equations which are suitable for the analysis of complex bifurcations which may occur in nonlinear engineering systems. We are particularly concerned with flow induced oscillations which occur in, for example, galloping transmission lines or panel flutter and related vibration problems.

We shall present a general framework for the analysis of these problems with the aim of extracting qualitative information, such as the existence and number of periodic orbits or rest points and their stability. This analysis is meant to complement existing techniques

---

*Present address: Department of Theoretical and Applied Mathematics, Cornell University.

such as asymptotic or numerical methods.

The present method uses intrinsically qualitative techniques, such as those of blowing up a singularity and of invariant manifolds. These approaches are particularly powerful for multiparameter problems.

After some dynamical preliminaries, we illustrate the technique of blowing up a singularity for bifurcation of fixed points and then pass to dynamic bifurcations. The last section discusses applications to various engineering systems exhibiting flutter, and in particular the problem of panel flutter.

§1. SOME PRELIMINARIES

In many problems concerning the bifurcation of equilibrium states, it is important to keep the full dynamical problem in mind. For example, stability is often best understood in the dynamical sense; also it may be useful to know that the bifurcated equilibria lie on an invariant manifold of low dimension for the full dynamical problem. Of course, if the bifurcations include oscillations (periodic orbits), it is impossible to ignore the dynamics.

We shall be interested in methods which are applicable to multiparameter systems. Indeed, this is often necessary to produce bifurcation diagrams which are insensitive to small perturbations in the equations. (In the literature this is variously studied under the headings "Perturbed Bifurcation Theory" - cf. Keener and Keller [26] or "Catastrophe Theory" - cf. Arnold [2] and Thom [55]. The most famous example of this is Euler buckling. Viewed as a one parameter system with parameter the beam tension, one gets the traditional picture shown in figure 1.

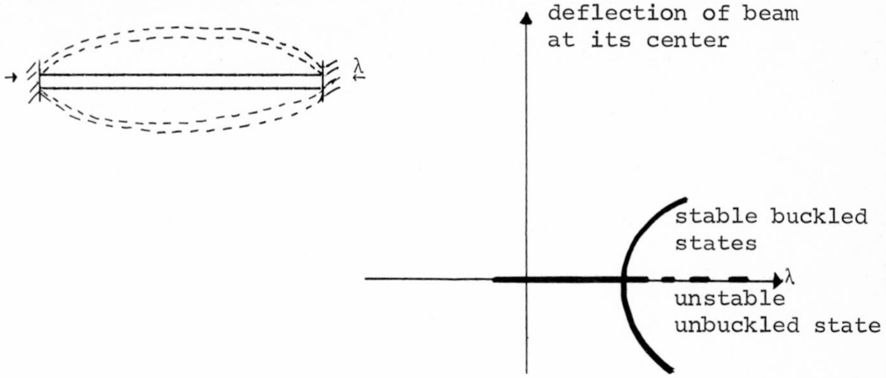

Figure 1

However, this bifurcation diagram is "unstable". It can be stablized by adding a second parameter ϵ , which describes the asymmetry of the force λ .† Now we get the more comprehensive bifurcation diagram shown in Figure 2. That in figure 1 is obtained by taking the slice $\epsilon = 0$. This new two-parameter bifurcation diagram is now qualitatively insensitive to further perturbations, since the cusp singularity is "structurally stable" [55].

---

†Bifurcations of the fixed points of Duffing's equation $\ddot{x} + \alpha\dot{x} + \gamma x^2\dot{x} - \lambda x + \delta x^3 + \epsilon = 0$ provide a model for this system. See Holmes and Rand [21] for a complete account.

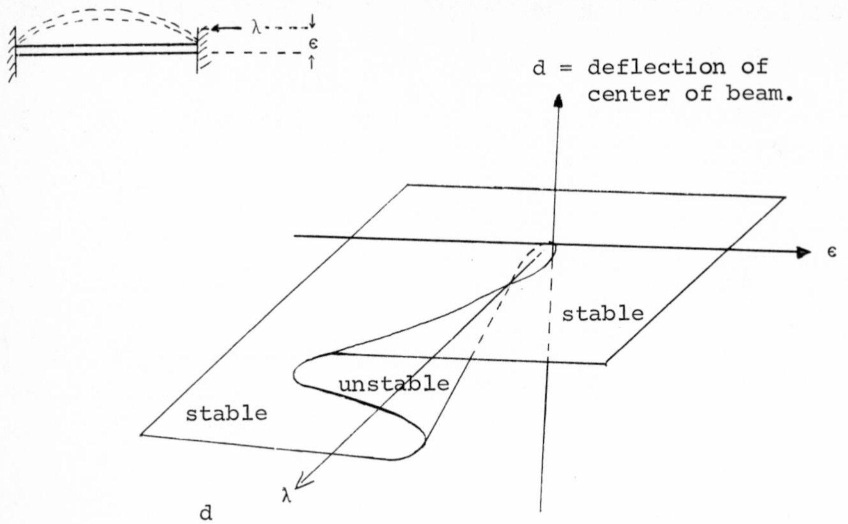

λ
ε
d = deflection of center of beam.
ε
stable
unstable
stable
λ

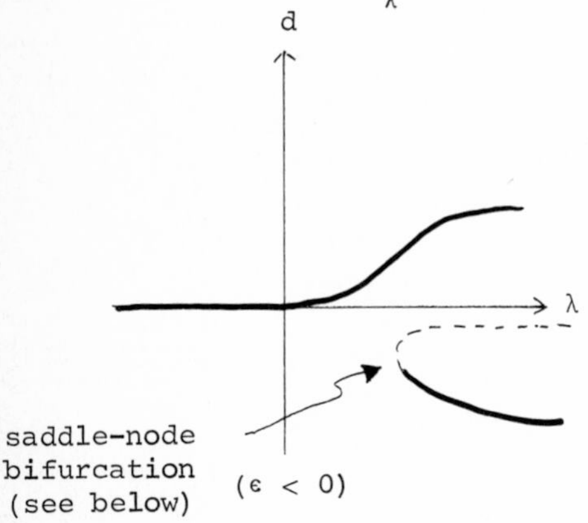

d
λ
saddle-node bifurcation (see below)
(ε < 0)

Figure 2.

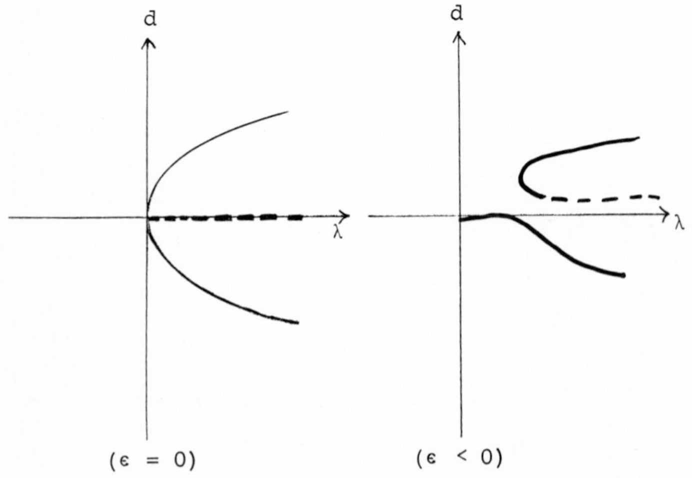

d
λ
(ε = 0)
d
λ
(ε < 0)

The reader can find full discussions of this kind of phenomenon from both the engineering and the mathematical point of view in Zeeman [59], Chow, Hale and Mallet-Paret [8], Thompson and Hunt [56] and Roorda [41-43], and references therein.

§1.1 THE MATHEMATICAL FRAMEWORK

The dynamical framework in which we operate is described as follows. Let $X \subset Y$ be Banach spaces (or manifolds) and let

$$f : X \times \mathbb{R}^p \to Y$$

be a given C^k mapping $(k \geq 2)$. Here $\mathbb{R}^p$ is the parameter space and f may be defined only on an open subset of $X \times \mathbb{R}^p$. The dynamics is given by

$$\frac{dx}{dt} = f(x,\lambda)$$

which defines a semi-flow

$$F_t^\lambda : X \to X$$

by letting $F_t^\lambda(x_0)$ be the solution of $\dot{x} = f(x,\lambda)$ with initial condition $x(0) = x_0$. We assume that this equation defines a local semi-flow on X; i.e. it has, at least locally in time, unique solutions.

A <u>fixed point</u> is point (x_0,λ) such that $f(x_0,\lambda) = 0$. Therefore, $F_t^\lambda(x_0) = x_0$ i.e. x_0 is an equilibrium point of the dynamics.

A fixed point (x_0,λ) is called <u>stable</u> if there is a neighborhood U_0 of x_0 on which $F_t^\lambda(x)$ is defined for all $t \geq 0$ and if for any neighborhood $U \subset U_0$, there is a neighborhood $V \subset U_0$ such that $F_t^\lambda(x) \in U$ if $x \in V$ and $t \geq 0$. The fixed point is called <u>asymptotically stable</u> if, in addition, $F_t^\lambda(x) \to x_0$ as $t \to +\infty$, for x in a neighborhood of x_0.

Many nonlinear partial differential equations of evolution type fall into this framework, as we shall see in §4. Also, many semi-linear hyperbolic and most parabolic type equations satisfy an additional smoothness condition; we say F_t^λ is a <u>smooth semi-flow</u> if for each t, λ, $F_t^\lambda : X \to X$ (where defined) is a C^k map and its derivatives are strongly continuous in t, λ.

For general conditions under which a semi-flow is smooth, see Marsden and McCracken [32]. One especially simple case occurs when

$$f(x,\lambda) = A_\lambda x + B(x,\lambda) ,$$

where $A_\lambda : X \to Y$ is a linear generator depending continuously on λ and $B : Y \times \mathbb{R}^p \to Y$ is a C^k map. This result is readily proved by the variation of constants formula

$$x(t) = e^{tA_\lambda} x_0 + \int_0^t e^{(t-s)A_\lambda} f(x(s),\lambda)\,ds$$

(See Segal [48] for details).

Standard estimates and the proof for ordinary differential

equations now prove the following (see Marsden and McCracken [32] for details):

LIAPUNOV'S THEOREM. *Suppose* F_t^λ *is a smooth flow,* (x_0,λ) *is a fixed point and the spectrum of the linear semi-group*

$$U_t^\lambda = D_x F_t^\lambda(x_0) \; : X \to X$$

(*The Fréchet derivative with respect to* $x \in X$) *is* $e^{t\sigma}$ *where* σ *lies in the left half plane a distance* $> \delta > 0$ *from the imaginary axis. Then* x_0 *is asymptotically stable and for* x *sufficiently close to* x_0 *we have an estimate*

$$\|F_t^\lambda(x) - x_0\| \leq Ce^{-t\delta} \quad .$$

If we are interested in the location of fixed points, we solve the equation

$$f(x,\lambda) = 0 \; ,$$

and the stability of a fixed point x_0 will be determined by the spectrum σ of the linearization at x_0 :

$$A_\lambda = D_x f(x_0,\lambda) \; .$$

(We assume the operator is non-pathological --- eg has discrete spectrum --- so $\sigma(e^{tA_\lambda}) = e^{t\sigma(A_\lambda)}$.) In critical cases where the spectrum lies on the imaginary axis, stability has to be determined by other means. It is at criticality where, for example, a curve of fixed points $x_0(\lambda)$ changes from stable to unstable, that a bifurcation can occur, as we shall see in §2.

The second major point we wish to make is that within the context of smooth semi-flows, the usual invariant manifold theorems from ordinary differential equations carry over.

In bifurcation theory it is often useful to apply the invariant manifold theorems to the <u>suspended flow</u>

$$\begin{aligned} F_t : X \times \mathbb{R}^p &\to X \times \mathbb{R}^p \\ (x,\lambda) &\to (F_t^\lambda(x),\lambda) \end{aligned}$$

The invariant manifold theorem states that if the spectrum of the linearization A_λ at a fixed point (x_0,λ) splits into $\sigma_S \cup \sigma_C$, where σ_S lies in the left half plane and σ_C is on the imaginary axis, then the flow F_t leaves invariant manifolds M_S and M_C tangent to the eigenspaces corresponding to σ_S and σ_C respectively; M_S is the <u>stable</u> and M_C is the <u>center</u> manifold. (One can allow an <u>unstable</u> manifold too if that part of the spectrum is finite). By Liapunov's theorem, orbits on M_S converge to (x_0,λ) exponentially. For suspended systems, note that we always have $1 \in \sigma_C$.

The idea of the proof is this: we apply invariant manifold theorems for smooth maps with a fixed point to each F_t separately. Since F_t and F_s commute $(F_t \circ F_s = F_{t+s} = F_s \circ F_t)$, it follows that these invariant manifolds can be chosen in common for all the

F_t.

For bifurcation problems the center manifold theorem is the most relevant, so we summarize the situation. (See Marsden and McCracken [32] for details).

CENTER MANIFOLD THEOREM FOR FLOWS. Let Z be a Banach space admitting a C^∞ norm away from 0 and let F_t be a C^0 semi-flow defined on a neighborhood of 0 for $0 \leq t \leq \tau$. Assume $F_t(0) = 0$ and for each $t > 0$, $F_t : Z \to Z$ is a C^{k+1} map whose derivatives are strongly continuous in t. Assume that the spectrum of the linear semigroup $DF_t(0) : Z \to Z$ is of the form $e^{t(\sigma_s \cup \sigma_c)}$ where $e^{t\sigma_c}$ lies on the unit circle (i.e. σ_c lies on the imaginary axis) and $e^{t\sigma_s}$ lies inside the unit circle a nonzero distance from it, for $t > 0$; i.e. σ_s is in the left half plane. Let Y be the generalized eigenspace corresponding to the part of the spectrum on the unit circle. Assume $\dim Y = d < \infty$.

Then there exists a neighborhood V of 0 in Z and a C^k submanifold $M_c \subset V$ of dimension d passing through 0 and tangent to Y at 0 such that

(a) If $x \in M_c$, $t > 0$ and $F_t(x) \in V$, then $F_t(x) \in M_c$.

(b) If $t > 0$ and $F_t^n(x)$ remains defined and in V for all $n = 0, 1, 2, \ldots$, then $F_t^n(x) \to M_c$ as $n \to \infty$.

See Figure 3 for a sketch of the situation.

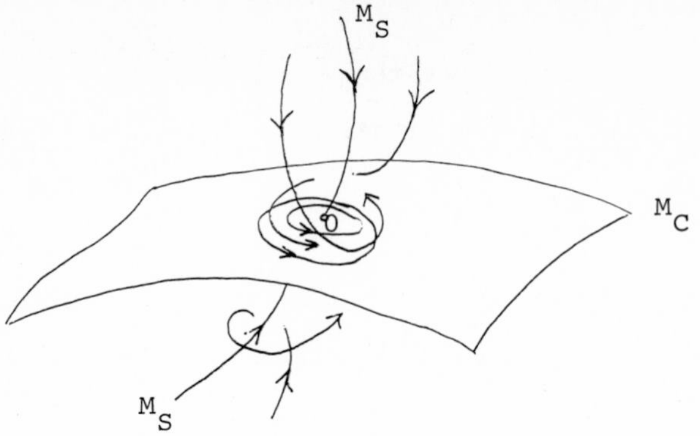

Figure 3

For example, suppose we have a curve of fixed points $x_0(\lambda)$, $\lambda \in \mathbb{R}$ which become unstable as λ crosses λ_0 and two stable fixed points branch off, as in figure 1. Then all three points will lie on the center manifold for the suspended system. Taking $\lambda =$ constant slices yields an invariant manifold M_C^{λ} for the parametrized system; see figure 4.

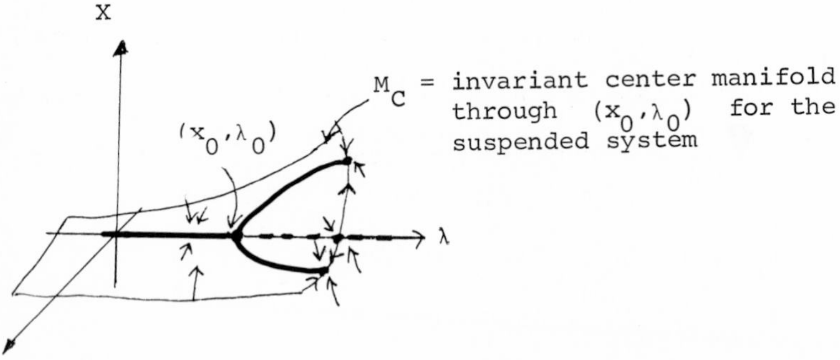

Figure 4 (also, see p. 170)

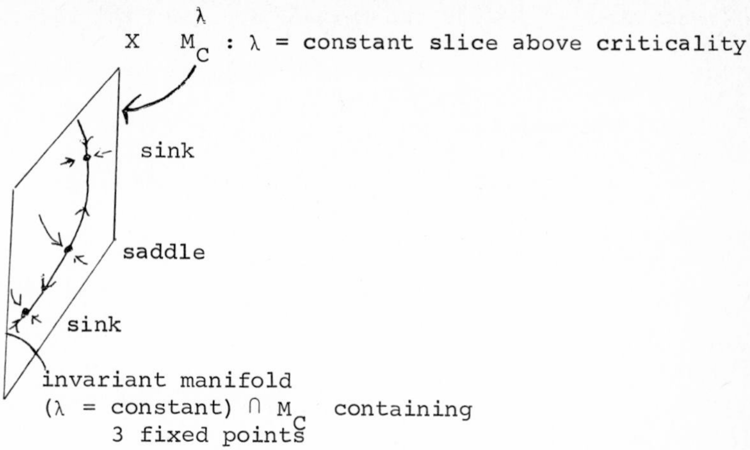

Figure 4 (cont'd)

Although the center manifold is only known implicity, it can greatly simplify the problem qualitatively by reducing an initially inifinite dimensional problem to a finite dimensional one. Likewise, questions of stability become questions on the center manifold itself. For example, it becomes clear, at least under a non-degneracy condition, that in the context of figure 4, supercritical branches are stable and subcritical branches are unstable. (For center manifolds of higher dimension, however, this is not true in general - see McLeod and Sattinger [30] for instance and the discussion in §3).

§2. BIFURCATION OF FIXED POINTS

Most of the literature on bifurcation theory deals with bifurcation of fixed points. For example, see Cesari [7], Sather [45, 46], Crandall and Rabinowitz [9, 10], Nirenberg [35], Sattinger

[47], articles in Keller and Antman [25], and the references therein. On the applications side, many of the papers grew out of work of Koiter (see Thompson and Hunt [56] for references).

Here we shall give a relatively simple geometrical framework for dealing with bifurcations at multiple eigenvalues for multiparameter systems. The approach follows Buchner, Marsden and Schecter [6] and combines some ideas in Nirenberg [35] with the method of blowing up a singularity. A general stability analysis is complex, as indicated in McLeod and Sattinger [30]. In specific problems this can sometimes be reduced to that for single parameter systems or an eigenvalue analysis can be done numerically, as we shall indicate in §4 . Therefore, we shall not discuss stability at this point any further.

As above, fixed points are determined by the zeros of a C^k map

$$f : X \times \mathbb{R}^p \to Y$$

Let $x_0(\lambda)$ be a given p-parameter manifold of solutions of $f(x,\lambda) = 0$ i.e. $f(x_0(\lambda),\lambda) = 0$ for λ in an open set in $\mathbb{R}^p$. Let $x_0(\lambda_0) = x_0$. Following standard terminology, we say that (x_0,λ_0) is a <u>bifurcation point</u> if every neighborhood of (x_0,λ_0) contains a solution (x,λ) of $f(x,\lambda) = 0$ with $x \neq x_0(\lambda)$. The set of all solutions near (x_0,λ_0) , including $(x_0(\lambda),\lambda)$, constitute the <u>bifurcation set</u>. From a more general point of view, it seems desirable to define a bifurcation point as one near which the set of solutions changes topological type as λ varies.

If (x_0,λ_0) is a bifurcation point, then $D_xf(x_0,\lambda_0) : X \to Y$, the Fréchet derivative of f with respect to x at (x_0,λ_0), is not a surjection. This is a trivial consequence of the implicit function theorem. Nevertheless, this criterion is effective in singling out condidates for bifurcation points.

Let $X_1 = \ker D_x f(x_0,\lambda_0)$ and assume

(i) X_1 splits; i.e. $X = X_1 \oplus X_2$ for a closed subspace $X_2 \subset X$

and (ii) X_1 is finite dimensional.

We refer to Buchner, Marsden and Schecter [6] for the case in which X_1 is allowed to be infinite dimensional.

Likewise, assume

(iii) Range $D_x f(x_0,y_0) = Y_1$ is closed and has a closed complement Y_2 ; $Y = Y_1 \oplus Y_2$, $\dim Y_2 < \infty$.

Let P be the projection of Y to Y_1 and let $x_2 = u(x_1,\lambda)$ be the unique solution of

$$f(x_1 + x_2, \lambda) = 0$$

for $x_1 \in X_1$, $x_2 \in X_2$ near (x_0,λ_0). This is guaranteed by the implicit function theorem. Thus, the equation $f(x,\lambda) = 0$ is equivalent to the <u>bifurcation equation</u>:

$$(I - P)f(x_1 + u(x_1,\lambda),\lambda) = 0$$

The reduction to the bifurcation equation, called the Liapunov-Schmidt procedure, is analogous to the reduction to the center manifold. In fact, as described above, the bifurcation of fixed points takes place within a center manifold for the dynamical systems.

Usually, but not always, $D_x f$ is a Fredholm map and so the bifurcation equation is a finite dimensional problem. (An exception occurs in general relativity; see Fischer, Marsden and Moncrief [13].) The methods of Buchner, Marsden and Schecter [6] do not require this assumption.

Now we give the main result on bifurcation at simple eigenvalues. (The hypotheses are stated in a form convenient for verification and are weakened below with no change in the proof). The result is essentially the same as in Crandall and Rabinowitz [9], to which the reader is referred for examples. The proof, however, is more geometrically satisfying. If is due to Nirenberg [35], based on a suggestion of Duistermaat.

THEOREM (BIFURCATION AT SIMPLE EIGENVALUES). <u>Assume</u> $p = 1$, <u>dim</u> X_1 = <u>dim</u> $Y_2 = 1$,

$$\frac{\partial f}{\partial \lambda}(x_0, \lambda_0) = 0, \quad \frac{\partial^2 f}{\partial \lambda^2}(x_0, \lambda_0) \in Y_1$$

<u>and</u> $\frac{\partial^2 f}{\partial \lambda \partial x}(x_0, \lambda_0) \cdot x_1 \notin X_1$ <u>where</u> X_1 = <u>span</u> (x_1), $\|x_1\| = 1$

<u>Then the bifurcation set near</u> (x_0, λ_0) <u>consists of two intersecting, transversal,</u> C^{k-2} <u>curves</u>.

PROOF. Let ℓ be a linear functional orthogonal to Y_1 and let

$$\varphi : X_1 \times \mathbb{R} \to \mathbb{R},$$

$$\varphi(x_1, \lambda) = \ell(f(x_1 + u(x_1, \lambda), \lambda)$$

so that, $\varphi^{-1}(0)$ is the bifurcation set near (x_0, λ_0). Easy calculations show that

$$\varphi(x_0, \lambda_0) = 0$$

$$d\varphi(x_0, \lambda_0) = 0$$

and

$$d^2\varphi(x_0, \lambda_0) = \begin{pmatrix} * & \ell\left(\frac{\partial^2 f}{\partial \lambda \partial x}(x_0, \lambda_0) \cdot x_1\right) \\ \ell\left(\frac{\partial^2 f}{\partial \lambda \partial x}(x_0, \lambda_0) \cdot x_1\right) & 0 \end{pmatrix}$$

Thus (x_0, λ_0) is a non-degenerate critical point for φ of index 1. Thus by a C^{k-2} change of coordinates, $\varphi(y,\mu) = \frac{1}{2}(y^2 - \mu^2)$ by the Morse lemma. Hence $\varphi^{-1}(0)$ is two C^{k-2} curves. □

With this point of view, we can now turn to the general case. We shall need a preliminary definition.

DEFINITION. Let Z_1, Y_2 be Banach spaces and $B : Z_1 \times Z_1 \to Y_2$ be a continuous symmetric bilinear form. Let $Q(v) = \frac{1}{2}B(v,v)$ be the associated quadratic form. Let $C = Q^{-1}(0)$; i.e. C is the cone of zeros of Q. We say Q is in general position at $v \in C$, $v \neq 0$ if the linear map $w \mapsto B(v,w)$ of Z_1 to Y_2 is surjective. If we say Q is in general position on C we mean it is so at each non-zero point of C.

THEOREM (THE GENERAL CASE). Let $f : X \times \mathbb{R}^p \to Y$ be as above. Assume (i), (ii) and (iii) hold, and that

$$\frac{\partial f}{\partial \lambda}(x_0, \lambda_0) = 0 .$$

Let $B = (I - P)D^2 f(x_0, \lambda_0)$ restricted to $X_1 \times \mathbb{R}^p = Z_1$ and Q be the associated quadratic form. Assume that Q is in general position on C.

Then the bifurcation set near (x_0, λ_0) is homeomorphic to $C = Q^{-1}(0)$ via a homeomorphism that takes (x_0, λ_0) to 0 and is a C^k diffeomorphism away from (x_0, λ_0).

If $v \in Q^{-1}(0)$, there is a C^{k-2} curve $(x(s), \lambda(s))$ of solutions to $f(x,\lambda) = 0$ tangent to v at (x_0, λ_0) and the union of these curves constitutes the bifurcation set.

REMARKS. 1. If one only knows Q is in general position at a particular $v \in C$, then v is still tangent to a C^{k-2} curve in the bifurcation set.

2. The proof may enable one to determine the structure of the bifurcation set even if the hypotheses fail. One may have to rescale the variables by different amounts in different directions and follow the method outlined in the blowing up lemma below.

3. If $p-1 + \dim X_1 = \dim Y_2 = m$, the bifurcation set consists on $2s$ curves of class C^{k-2} through (x_0, λ_0) where $1 \leq s \leq 2^{m-1}$ (This follows by using Bezout's theorem from algebraic geometry to determine the number of rays in $Q^{-1}(0)$).

The theorem is proved by appeal to a rather general result. The interest in this approach is that the techniques are completely straightforward and applicable to a wide variety of situations. We use the following lemma:

BLOWING-UP LEMMA. *Let H be Euclidean n-space, V Euclidean m-space and $g : H \to Y$ a C^k map, $k \geq 3$. Assume*

$$Q \text{ is in general position on } C = Q^{-1}(0)$$

Then there is a neighborhood U of 0 in H such that $g^{-1}(0) \cap U$ is homeomorphic to $Q^{-1}(0)$ via a homeomorphism that takes 0 to 0 and is a C^k diffeomorphism away from 0. Moreover, if $v \in Q^{-1}(0)$, there is a C^{k-2} curve $\alpha(s) \in g^{-1}(0)$, $-\delta \leq s \leq \delta$ with $\alpha(0) = 0$, $\alpha'(0) = v$.

Here is how the lemma yields the theorem. Let $g : H = X_1 \times \mathbb{R}^p \to Y_2 = V$ be defined by

$$g(x_1, \lambda) = (I - P) f(x_1 + u(x_1, \lambda), \lambda)$$

Clearly g is of class C^k. Also, noting that $Du(x_0, \lambda_0) = 0$ (from the definition of u and $\frac{\partial f}{\partial \lambda}(x_0, \lambda_0) = 0$), one calculates that

(a) $Dg(x_{10},\lambda_0) = 0$

and (b) $D^2g(x_{10},\lambda_0) = (I - P)\, D^2f(x_0,\lambda_0)$, restricted to $X_1 \times \mathbb{R}^p$.

Therefore, by our assumptions in the main theorem, the blowing-up lemma applies to g . Since the zeros of $f(x,\lambda)$ are the graph of u over the zeros of g , the conclusions of the theorem follow.

Here is the idea of the proof of the blowing-up lemma. Let S be the unit sphere in H . Set

$$\tilde{g} : S \times \mathbb{R} \rightarrow V ,$$

$$\tilde{g}(x,r) = \frac{1}{r^2}\, g(rx)$$

By Taylor's theorem,

$$g(x) = Q(x) + R(x)$$

so where R is C^{k-2} , so

$$\tilde{g}(x,r) = Q(x) + \frac{1}{r^2} R(rx)$$

and since R vanishes like r^3 , $\tilde{g}$ is C^{k-2} . Away from $r = 0$, the zeros of $\tilde{g}$ and g are in 1-1 correspondence. If we identify $S \times \{0\}$ and S , we have thus blown up the singularity of g at 0 to the unit sphere S . Near S , the structure of the set of zeros of g can be analyzed easily since 0 is a regular value of g on S (by hypothesis) and $\tilde{g}^{-1}(0)$ intersects S transversally. By pushing this structure down to X_1 by the map $(x,r) \mapsto rx$, we get the result.

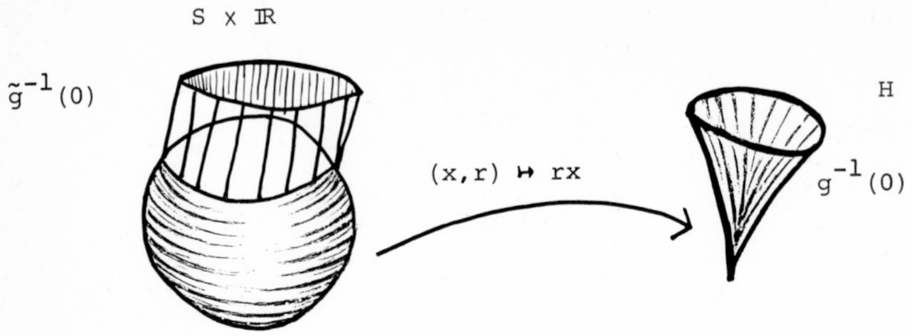

Figure 5.

§3. DYNAMIC BIFURCATION THEORY

Bifurcation theory for dynamical systems is much less developed than that for fixed points. Indeed the variety of bifurcation possible and their structure is much more complex. We shall briefly outline here some examples of dynamic bifurcations and then state a general plan for attacking a complex bifurcation problem.

We begin by describing the simplest bifurcations for <u>one</u> parameter systems. In a sense these bifurcations are the generic local ones. (See Sotomayor [50] and Takens [53] for details) . If one imposes a symmetry, however, what is generic may change, as we shall explain.

SADDLE NODE. This is a bifurcation of fixed points; a saddle and a sink come together and annihilate one another, as shown in Figure 6 . A simple real eigenvalue of the sink crosses the imaginary axis at the moment of bifurcation; one for the saddle crosses in the opposite direction. The suspended center manifold is 2-dimensional . The symmetric situation of a saddle-source is also possible.

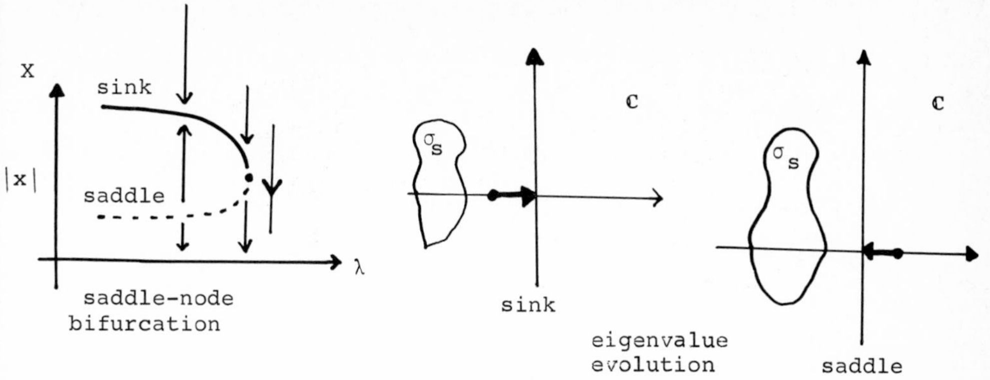

Figure 6.

If an axis of symmetry is present, as will be the case in the example of panel flutter treated in §4 , then a symmetric bifurcation can occur, as in Figure 7 . As in our discussion of Euler buckling, a small asymmetric perturbation or imperfection 'unfolds' this into a simple non bifurcation path and a saddle node. In Figures 6 and 7 we also indicate the vector field flow directions schematically.

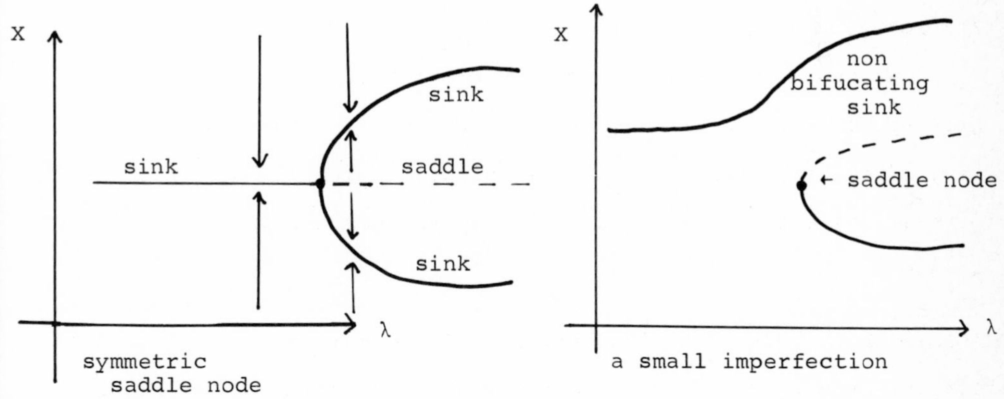

Figure 7.

HOPF BIFURCATION. This is a bifurcation to a periodic orbit; here a sink becomes a saddle by two complex conjugate non-real eigenvalues crossing the imaginary axis. As with the symmetric saddle node, the bifurcation can be sub-(unstable closed orbits) or super-(stable closed orbits) critical. (See Marsden and McCracken [32] for calculations to determine which is which) . Figure 8 depicts the supercritical attracting case. Here the suspended center manifold is 3-dimensional.

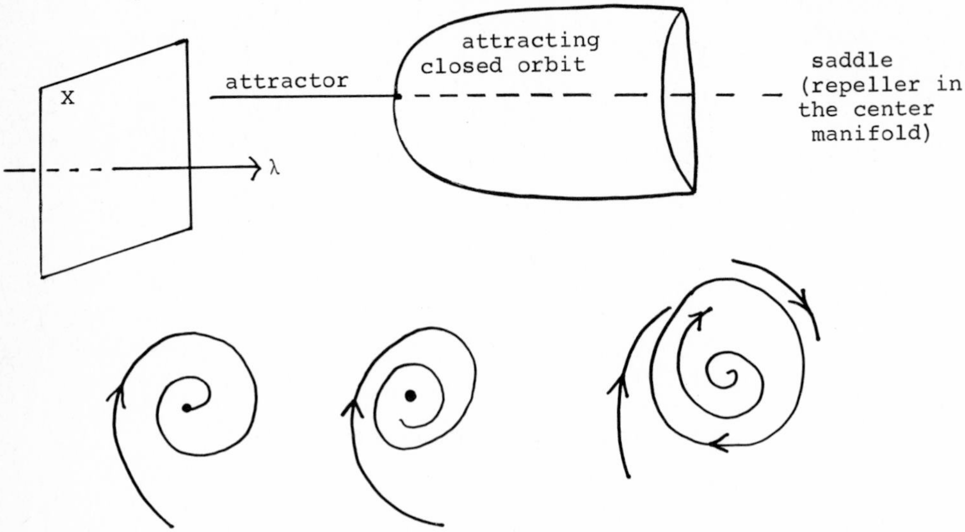

Figure 8 The Hopf Bifurcation.

These two bifurcations are local in the sense that they can be analyzed by linearization about a fixed point. There are however, some global bifurcations which are more difficult to detect. A saddle connection is shown in Figure 9; cf Takens [54] , Arnold [2] .

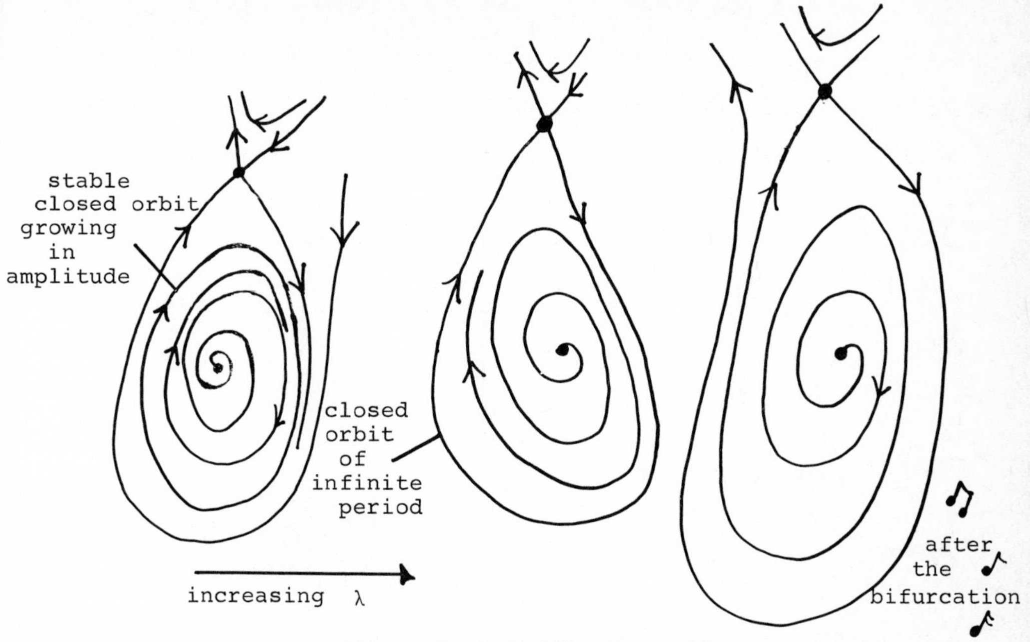

Figure 9. A Saddle Connection

Here the stable and unstable seperatrices of the saddle point pass through a state of tangency (when they are identical) and thus cause the annihilation of the attracting closed orbit.

These global bifurcations can occur as part of local bifurcations of systems with additional parameters. This approach has been developed by Takens [54] who has classified generic or 'stable' bifurcations of two parameter families of vector fields on the plane. This is an outgrowth of extensive work of the Russian school led by Andronov [1] . An example of one of Taken's bifurcations with a symmetry imposed is shown in Figure 10. (The labels are for later use.)

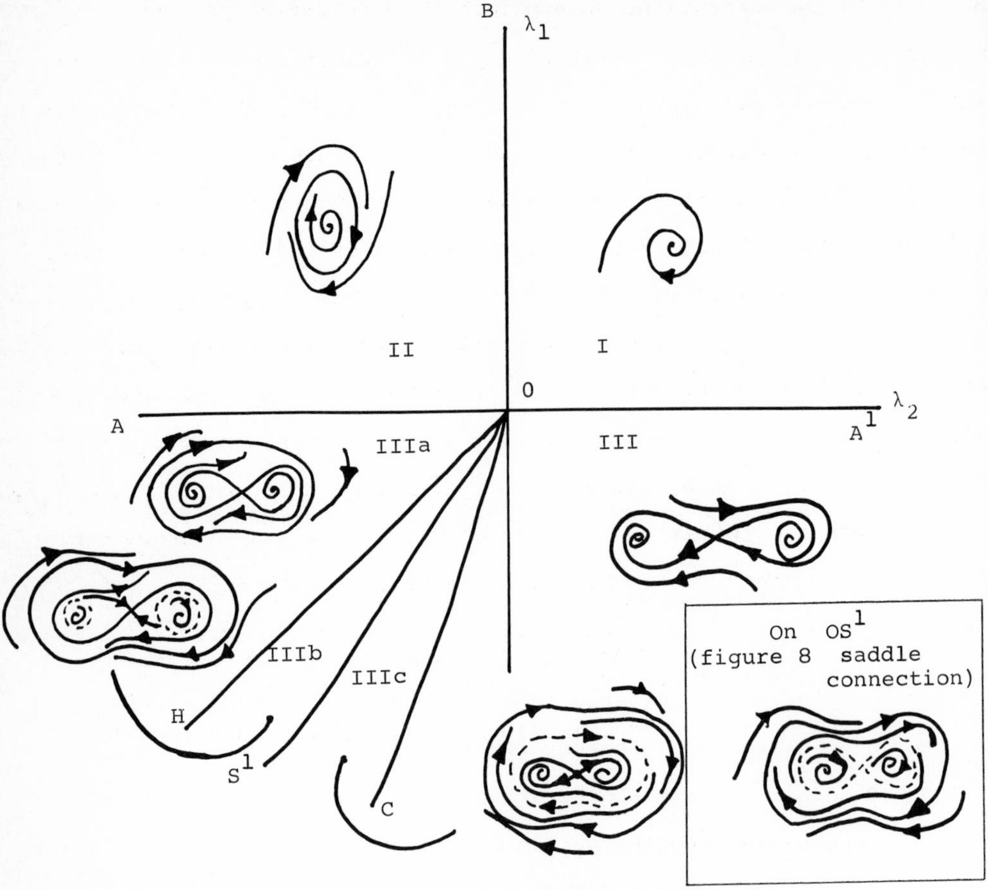

Figure 10. Takens' (2,-) normal form showing the local phase portrait in each region on parameter space (Takens' [54])

Some of the phenomena captured by the bifurcations outlined above have been known to engineers for many years. In particular we might mention the jump phenomenon of Duffings equation (see Timoshenko [57], Holmes and Rand [21]) and the more complex bifurcational behavior of the forced van der Pol oscillator (Hayashi [15], Holmes and Rand [22] ; [22] contains a proof that the planar variational equation of the latter oscillator undergoes a saddle connection bifurcation as in Figure 9 .

Now we can outline an approach to bifurcation problems (cf. Holmes and Rand [20]) . First of all, the analysis is for two parameter systems which posess, near a fixed point (x_0, λ_0), a three dimensional suspended center manifold (i.e. for each λ , a two dimensional invariant manifold for the dynamics). Typically, a fixed point will have a real double zero eigenvalue at a certain parameter value and we are interested in bifurcations near this organizing center. One first fills in as much of the bifurcation diagram as possible, using linearization to detect Hopf and saddle node bifurcation. Second, one assumes (taking any symmetry into account) that the bifurcation diagram itself is stable to small perturbations. This is justified since one is presumably working with a model which only approximates some physical situation [20] . Finally, the correct bifurcation diagram is obtained by looking through Takens' list for a diagram(s)† consistent with the information obtained.

We shall illustrate how this procedure works in a concrete problem in §4 .

§4. FLUTTER IN ENGINEERING SYSTEMS

Before giving a particular example analyzed by the methods of §3 , we discuss some ideas and examples of flutter in general.

---

† See Takens [54] . Here, in §§5,6, Takens lists generic bifurcations of 2 parameter vectorfields on the plane (or on two-manifolds) having singularities with double zero eigenvalues. He allows the vectorfields to have rotational symmetry but assumes that there is no "higher" degeneracy in the nonlinear terms. See Takens [51] for an example where the latter does occur. It is not strictly correct to speak of a "list" of two-parameter bifurcation, since the various analyses has not been conveniently gathered in one article.

All too often, engineers are content with only a linear analysis. For example, flutter is often viewed as the presence of two complex conjugate eigenvalues with positive real part. (cf. Zeigler [60]). The non-linear system may be fluttering (i.e. have a closed orbit) or not, as shown in Figure 11 . Mathematically, the development of spontaneous flutter is best detected through the Hopf bifurcation, remembering that the periodic orbits could be unstable and the bifurcation subcritical.

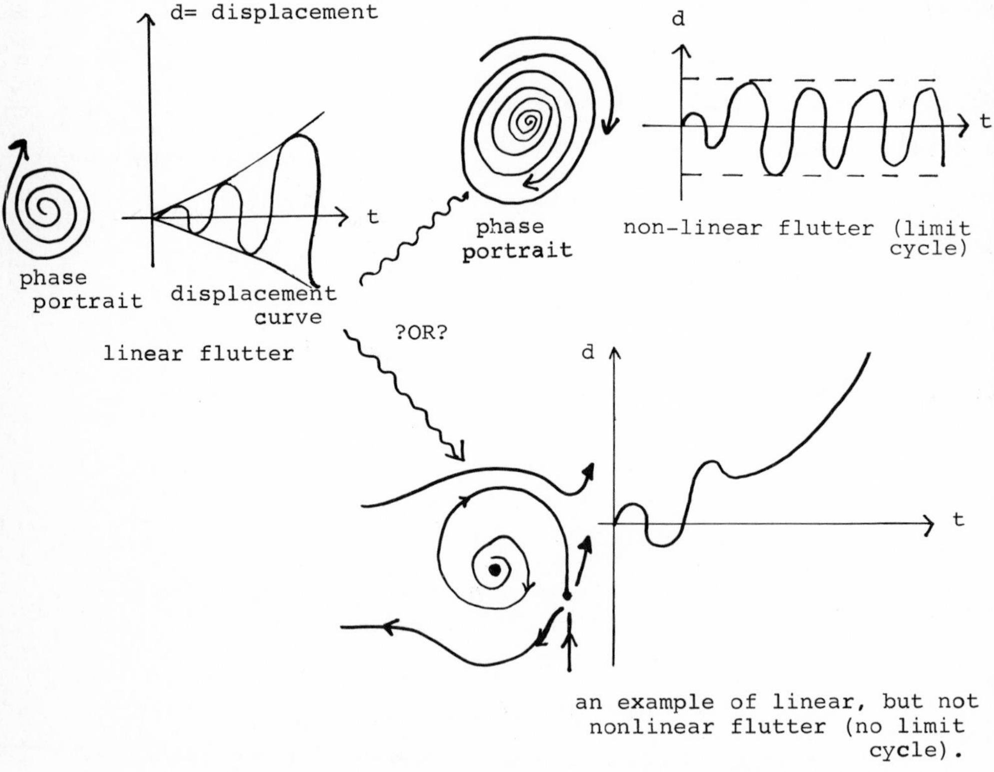

Figure 11.

Similar remarks may be made about divergence (a saddle point or source) as shown in Figure 12 .

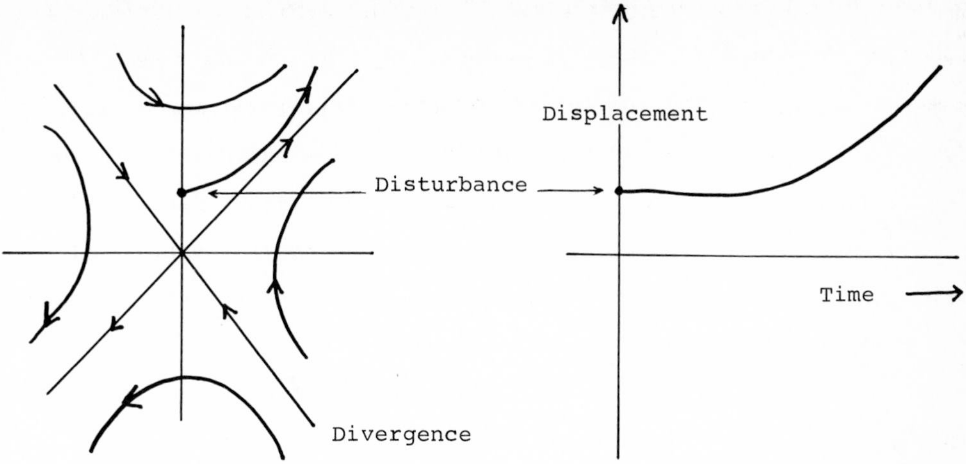

(a) linear theory

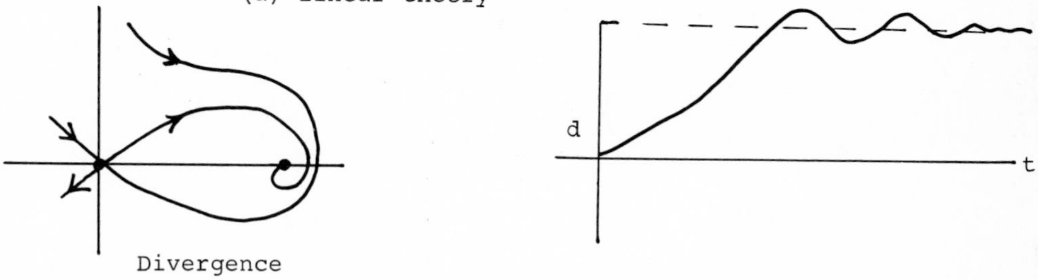

(b) a nonlinear possibility

Figure 12.

There are, in broad terms, three kinds of flutter of interest to the engineer. Here we briefly discuss these types. Our bibliography is not intended to be exhaustive, but merely to provide a starting point for the interested reader.

(a) AIRFOIL OR WHOLE WING FLUTTER ON AIRCRAFT

Here linear stability methods <u>do</u> seem appropriate since virtually any oscillations are catastrophic. Control surface flutter probably comes under this heading also. See Bisplinghoff and Ashley [3] and Fung [14] for examples and discussion.

(b) CROSS FLOW OSCILLATIONS

The familiar flutter of sun-blinds in a light wind comes under this heading. The "galloping" of power transmission lines and of tall buildings and suspension bridges provide examples which are of more direct concern to engineers: the famous Tacoma Narrows bridge disaster was caused by cross flow oscillations. In such cases (small) limit cycle oscillations are acceptable (indeed, they are inevitable), and so a nonlinear analysis is appropriate.

Cross-flow flutter is due to the oscillating force caused by "von-Karman" vortex shedding behind the body, Figure 13 .

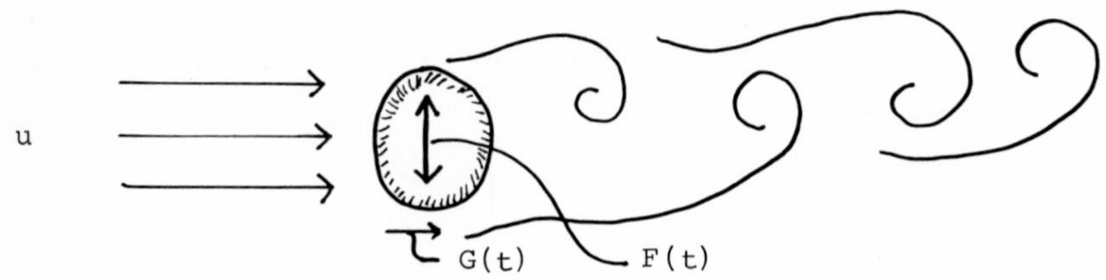

Figure 13. Cross flow oscillations

The alternating stream of vortices leads to an almost periodic force F(t) transverse to the flow in addition to the in-line force G(t) ; G(t) varies less strongly than F(t) . The flexible body responds to F(t) and, when the shedding frequency (a function of fluid velocity, u, and the body's dimensions) and the body's natural or resonance frequency are close, then "lock in" or entrainment can occur and large amplitude oscillations are observed. Experiments strongly suggest a limit cycle mechanism and engineers have traditionally modelled the situation by a van der Pol oscillator or perhaps a pair of coupled oscillators. See the symposium edited by Naudascher [34] for a number of good survey articles; the review by Parkinson is especially relevant. In a typical treatment, Novak [36] discusses a specific example in which the behaviour is modelled by a free van der Pol type oscillator with nonlinear damping terms of the form

$$a_1\dot{x} + a_2\dot{x}^2 + a_3\dot{x}^3 + \ldots$$

Such equations possess a fixed point at the origin $x = \dot{x} = 0$ and can also possess multiple stable and unstable limit cycles. These cycles are created in bifurcations as the parameters $a_1, a_2, \ldots,$ which contain windspeed terms, vary. Bifurcations involving the fixed point and global bifurcations in which pairs of limit cycles are created both occur (cf. Novak [36] figures 3,8,9). Parkinson also discusses the phenomenon of entrainment which can be modelled by the forced van der Pol oscillator.

In a more recent study, Landl [28] discusses such an example which displays both "hard" and "soft" excitation, or, in Arnold's term [2] , strong and weak bifurcations:

$$\ddot{x} + \delta\dot{x} + x = a\Omega^2 C_L$$

$$\ddot{C}_L + (\alpha - \beta C_L^2 + \gamma C_L^4)\dot{C}_L + \Omega^2 C_L = b\dot{x} .$$

Here $\cdot \equiv \frac{d}{dt}$ and $\alpha, \beta, \gamma, \delta, a, b$ are generally positive constants for a given problem (they depend upon structural dimensions, fluid properties, etc.). Ω is the vortex shedding frequency. As Ω varies the system can develop limit cycles leading to a periodic variation in C_L, the lift coefficient. The term $a\Omega^2 C_L$ then acts as a periodic driving force for the first equation, which represents one mode of vibration of the structure. This model, and that of Novak, appear to display generalised Hopf bifurcations (see Takens [51]) .

In related treatments allowance has been made for the effects of (broad band) turbulence in the fluid stream by including stochastic excitations. Vacaitis et. al [58] proposed such a model for the oscillations of a two degree of freedom structure and carried out some numerical and analogue computer studies. Recently Holmes and Lin [17] applied qualitative dynamical techniques to a deterministic version of this model prior to stochastic stability studies of the full model (Lin and Holmes [29]). The Vacaitis model assumes that the von Karman vortex excitation can be replaced by a term

$$F(t) \equiv F\cos(\Omega t + \Psi(t))$$

where Ω is the (approximate) vortex shedding frequency and $\Psi(t)$ is a random phase term. In common with all the treatments cited above the actual mechanism of vortex generation is ignored and "dummy" drag and lift coefficients are introduced. These provide discrete analogues of the actual fluid forces on the body. Iwan and Blevins [24] and St. Hilaire [49] have gone a little further in attempting to relate such force coefficients to the fluid motion but the problem appears so difficult that a rigorous treatment is still impossible. The major problem is, of course, our present inability to solve the Navier-Stokes equations for viscous flow around a body. Potential flow solutions are of no help here, but recent advances in numerical techniques may be useful. Ideally a rigorous analysis of the fluid motion should be coupled with a continuum mechanical analysis of the structure. For the latter, see the elegant Hamiltonian formulation of Marietta [31] for example.

The common feature of all these treatments (with the exception of Marietta's) is the implicit reduction of an infinite dimensional problem to one of finite dimensions, generally to a simple nonlinear oscillator. The use of center manifold theory and the concepts of genericity and structural stability suggests a way in which this reduction might be rigorously justified. To illustrate this we turn to the third broad class of flutter, which we discuss in more detail.

(c) AXIAL FLOW INDUCED OSCILLATIONS

In this class of problems, oscillations are set up directly through the interaction between a fluid and a surface across which it is moving. Examples are oscillations in pipes and (supersonic) panel flutter. Experimental measurements (vibration records from nuclear reactor fuel pins, for example) indicate that axial flow induced oscillations present a problem just as severe as the more obvious one of cross flow oscillations. See the monograph by Dowell [11] for an account of panel flutter and for a wealth of further references. Oscillations of beams in axial flow and of pipes conveying fluid have been studied by Païdoussis [37,38] and Brooke-Benjamin [4,5] ; see Païdoussis [38] for a good survey. Figure 14 shows the three situations.

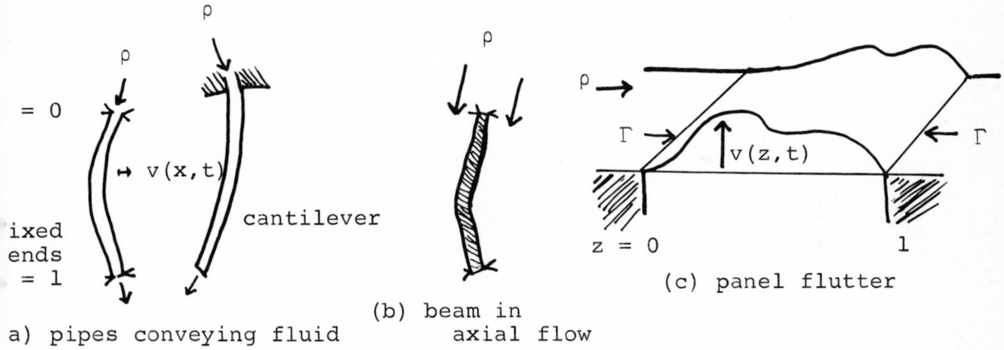

Figure 14. Axial flow-induced oscillations.

In addition to the effects of the fluid flow velocity ρ, the structural element might also be subject to mechanical tensile or compressive forces Γ which can lead to buckling instabilities even in the absence of fluid forces.

The equations of motion of such systems, written in one dimensional form and with all coefficients suitably nondimensionalised, can be shown to be of the type

$$\alpha\dot{v}'''' + v'''' - [K\int_0^1 (v'(\xi))^2 d\xi + \sigma\int_0^1 (v'(\xi)\dot{v}'(\xi))d\xi]v'' + \ddot{v} +$$

$$(*) \quad + [\text{linear fluid and mechanical loading terms in } v'', \dot{v}', v', \dot{v}] = 0$$

Here $\alpha, \sigma > 0$ are structural viscoelastic damping coefficients and $K > 0$ is a (nonlinear) measure of membrane stiffness; $v = v(z,t)$ and $\cdot = \partial/\partial t$; $' = \partial/\partial z$ (cf. Holmes [16]). Brooke-Benjamin [4] , Païdoussis [37,38] and Dowell [11,12] , for example, provide derivations of specific equations of this type. The fluid forces are again approximated, but in a more respectable manner.

In the case of panel flutter, if a static pressure differential exists across the panel, the right hand side carries an additional parameter P . Similarly, if mechanical imperfections exist so that compressive loads are not symmetric, then the "cubic" symmetry of (*) is destroyed (cf. §1, figures 1 and 2, above) .

Problems such as those of figure 14 have been widely studied both theoretically and experimentally, although, with the notable exception of Dowell and a number of other workers in the panel flutter area, engineers have concentrated on linear stability analyses. Such analyses can give misleading results, as we shall see. In many of these problems, engineers have also used low dimensional models, even though the full problem has infinitely many degrees of freedom. Such a procedure can actually be justified if careful use is made of the center manifold theorem.

Often the location of fixed points and the evolution of spectra about them has to be computed by making a Galerkin or other approximation and then using numerical techniques. There are obvious convergence problems (see Holmes and Marsden [18,19]), but once this is done, the organizing centers and dimension of the center manifolds can be determined relatively simply.

§4.1 PIPES CONVEYING FLUID AND SUPPORTED AT BOTH ENDS

Pipe flutter is an excellent illustration of the difference between the linear prediction of flutter and what actually happens in the PDE model. The phase portrait on the center manifold in the non-linear case is shown in figure 15, at parameter values for which the linear theory predicts "coupled mode" flutter. (cf. Païdoussis-Issid [38] and Plaut-Huseyin [40]) .

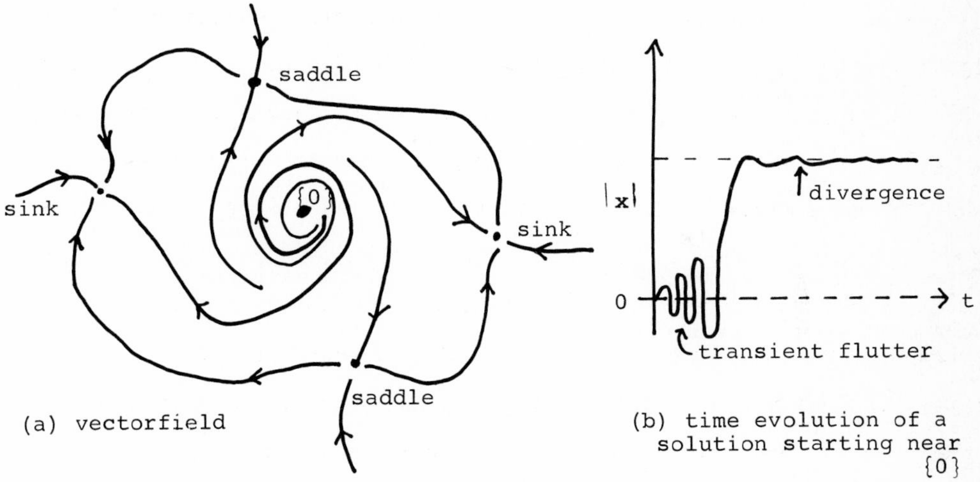

(a) vectorfield

(b) time evolution of a solution starting near {0}

Figure 15.

In fact, we see that the pipe merely settles to one of the stable buckled rest points with no non-linear flutter. (See Holmes [16] for details.) The presence of imperfections should not substantially change this situation (see Zeeman [59] for a global analysis of a similar two-mode buckling problem).

The absence of flutter in the nonlinear case can be seen by differentiating a suitable Liapunov function along solution curves of the PDE. In the pipe flutter case the PDE is

$$\alpha \dot{v}'''' + v'''' - \{\Gamma - \rho^2 + \gamma(1-z) + K|v'|^2 + \sigma(v',\dot{v}')\}v'' + $$
$$+ 2\sqrt{\beta}\rho\dot{v}' + \gamma v' + \delta\dot{v} + \ddot{v} = 0 \quad ,$$

(see Païdoussis-Issid [38], Holmes [16]). Here $|\cdot|$ and $(\cdot\,,\cdot)$ denote the usual L_2 norm and inner product and solutions $x = \{v, \dot{v}\}$ lie in a Hilbert space $X = H_0^2([0,1]) \times L_2([0,1])$ (see Holmes and Marsden [18,19] and §4.2 below for more details of the specific analytic framework for such a problem). For our Liapunov function we choose the energy: in this case given by

$$H(x(t)) = \tfrac{1}{2}|\dot{v}|^2 + \tfrac{1}{2}|v''|^2 + \frac{\Gamma-\rho^2}{2}|v'|^2 + \frac{K}{4}|v'|^4 + \frac{\gamma}{2}([1-z]v',v') \quad ,$$

(cf. Païdoussis-Issid [38], appendix I). Differentiating $H(x(t))$ along solution curves yields

$$\frac{dH}{dt} = -\delta|\dot{v}|^2 - \alpha|\dot{v}''|^2 - \sigma(v',\dot{v}')^2 - 2\sqrt{\beta}\rho(\dot{v}',\dot{v})$$

Since $(\dot{v}',\dot{v}) \equiv 0$ and $\delta, \alpha, \sigma > 0$, dH/dt is negative for all $v > 0$ and thus all solutions must approach rest points $x_i \in X$. In particular, for $\Gamma > \Gamma_0$, the first Euler buckling load, all solutions approach $x_0 = \{0\} \in X$ and the pipe remains straight. Thus a term of the type $\rho\dot{v}'$ cannot lead to nonlinear flutter. In the case of a beam in axial flow terms of this type <u>and</u> of the type $\rho^2 v'$ both occur and nonlinear flutter evidently can take place (see Païdoussis

[37] for a linear analysis). Experimental observations actually indicate that fluttering motions more complex than limit cycle can occur.

We should note, however, that <u>cantilevered pipes</u> can flutter: Brooke-Benjamin [5] has some excellent photographs of a two-link model. Here flutter is caused by the so-called follower force at the free end which introduces an additional term into the energy equation (see Brooke-Benjamin [4]).

§4.2 PANEL FLUTTER

Now we turn to an analysis of panel flutter. We consider the "one-dimensional" panel shown in Figure 14(c)† and we shall be interested in bifurcations near the trivial zero solution. The equation of motion of such a thin panel, fixed at both ends and undergoing "cylindrical" bending (so spanwise bending) can be written as

$$\alpha\dot{v}'''' + v'''' - \{\Gamma + K\int_0^1 (v'(\xi))^2 d\xi + \sigma\int_0^1 (v'(\xi)\dot{v}'(\xi))d\xi\}v'' +$$

$$+ \rho v'' + \sqrt{\rho}\,\delta\dot{v} + \ddot{v} = 0\ , \qquad (1)$$

see Dowell [12], Holmes [16]. Here $\cdot \equiv \partial/\partial t$, $' = \partial/\partial z$ and we have included viscoelastic structural damping terms α, σ as well as aerodynamic damping $\sqrt{\rho}\delta$. K represents nonlinear (membrane) stiffness, ρ the dynamic pressure and Γ an in-plane tensile load. All quantities are nondimensionalised and associated with (1) we have boundary conditions at $z = 0,1$ which might typically be simply

---

† A two dimensional or von-Karmen panel is presumably a good deal more complicated. For the bifurcation of fixed points, see Chow, Hale and Mallet-Paret [8].

supported $(v = v'' = 0)$ or clamped $(v = v' = 0)$. In the following we make the physically reasonable assumption that $\alpha, \sigma, \delta, K$ are fixed > 0 and let the control parameter $\mu = \{\rho, \Gamma | \rho \geq 0\}$ vary. In contrast to previous studies (Dowell [11,12]) in which (1) and similar equations were analyzed for specific parameter values and initial conditions by numerical integration of a finite dimensional Galerkin approximation, here we study the qualitative behavior of (1) under the action of μ.

To proceed with the methods of §3, we first redefine (1) as an ODE on a Banach space, choosing as our basic space $X = H_0^2([0,1]) \times L^2([0,1])$, where H_0^2 denotes H^2 functions† in $[0,1]$ which vanish at $0,1$. Set $\|\{v,\dot{v}\}\|_X = (|\dot{v}|^2 + |v''|^2)^{\frac{1}{2}}$, where $|\cdot|$ denotes the usual L^2 norm and define the linear operator

$$A_\mu = \begin{pmatrix} 0 & I \\ C_\mu & D_\mu \end{pmatrix} ; \qquad \begin{aligned} C_\mu v &= -v'''' + \Gamma v'' - \rho v' \\ D_\mu \dot{v} &= -\alpha \dot{v}'''' - \sqrt{\rho}\, \delta \dot{v} \end{aligned} . \tag{2}$$

The basic domain of A_μ, $D(A_\mu)$ consists of $\{v,\dot{v}\} \in X$ such that $\dot{v} \in H_0^2$ and $v + \alpha\dot{v} \in H^4$; particular boundary conditions necessitate further restrictions. After defining the nonlinear operator $B(v,\dot{v}) = (0, [K|v'|^2 + \sigma(v',\dot{v}')]v'')$, where $(\, , \,)$ denotes the L^2 inner product, (1) can be rewritten as

$$\frac{dx}{dt} = A_\mu x + B(x) \equiv G_\mu(x) ;$$

$$x = \{v,\dot{v}\}; \; x(t) \in D(A_\mu) . \tag{3}$$

---

† H^2 consists of functions which, together with their first and second derivative, are square integrable.

We next define an energy function $H : X \to \mathbb{R}$ by

$$H\{v,\dot{v}\} = \tfrac{1}{2}|\dot{v}|^2 + \tfrac{1}{2}|v''|^2 + \frac{\Gamma}{2}|v'|^2 + \frac{K}{4}|v'|^4 \tag{4}$$

and compute

$$\frac{dH}{dt} = -\rho(v',\dot{v}) - \sqrt{\rho}\,\delta|\dot{v}|^2 - \alpha|\dot{v}''|^2 - \sigma(v',\dot{v})^2 .$$

Using the methods of Segal [48] one shows that (3) and hence (1) defines a unique smooth local semi-flow F_t^{μ} on X. Using the energy function (4) and some arguments of Parks [39], one shows that $H(x(t))$ is bounded and hence that F_t^{μ} is in fact globally defined for all $t \geq 0$.

By making 2-mode and 4-mode approximations, one finds that for $\sigma = 0.0005$, $\delta = 0.1$, the operator A_{μ} has a double zero eigenvalue at $\mu = (\rho, \Gamma) \approx (110, -22.6)$, (the point 0 in figure 16) the remaining eigenvalues being in the left half plane. (See Holmes [16] and Holmes-Marsden [18, 19].) Thus around the zero solution we obtain a four dimensional[†] suspended center manifold. Referring to the eigenvalue evolution at the zero solution in Figure 17, which is obtained numerically, we are able to fill in the portions of the bifurcation diagram shown in Figure 16.

---

† Note that the control parameter μ is now two dimensional.

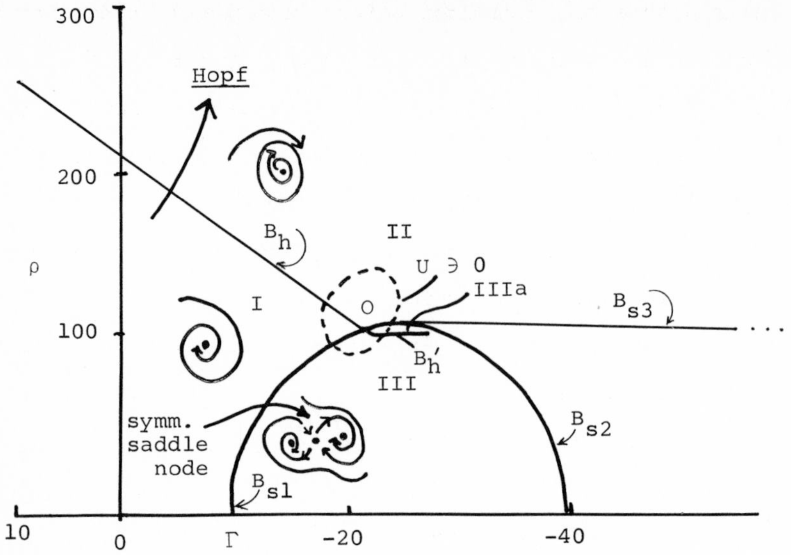

Figure 16. Partial bifurcation set for the two mode panel (α=0.005, δ= 0.1).

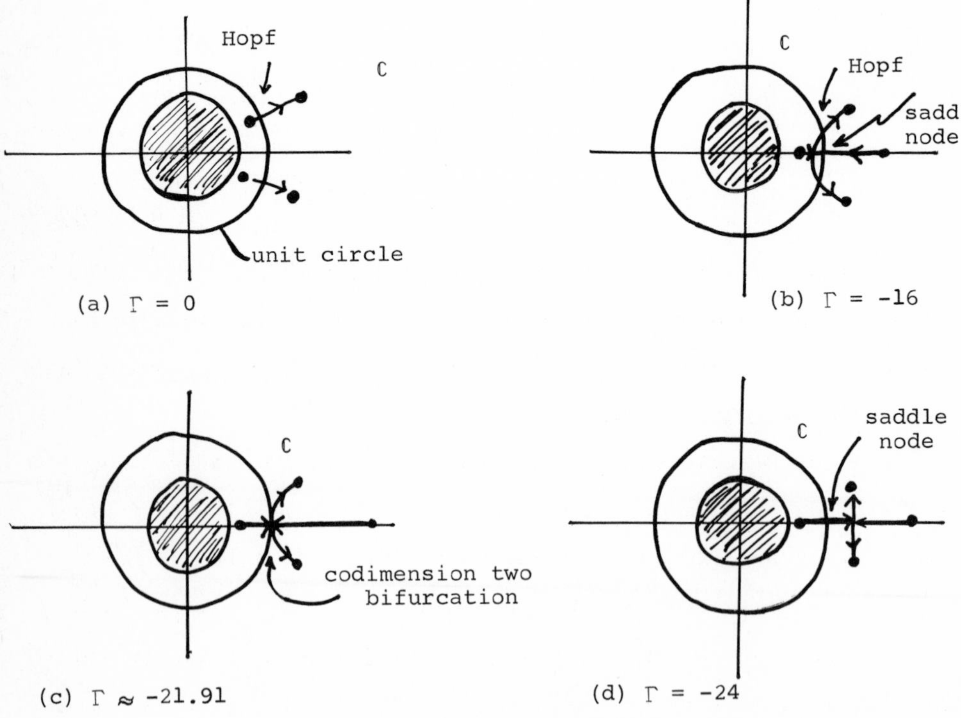

Figure 17. Eigenvalue evolutions for $DF_t^{\mu}(0) : X \to X$, Γ fixed, ρ increasing, estimated from the 2 mode model

In particular a supercritical Hopf bifurcation occurs crossing B_h and a symmetrical saddle node on B_{s1}, as shown. These are the flutter and buckling or divergence instabilities detected in previous studies such as Dowell's. Moreover, finite dimensional computations for the two fixed points $\{\pm x_0\}$ appearing on B_{s1} and existing in region III show that they are sinks ($|\text{spectrum}\ (DF_t^{\mu}(\pm x_0))| < 1$) below a curve B_h' originating at 0 which we also show on figure 16. As μ crosses B_h' transversally, $\{\pm x_0\}$ undergo simultaneous Hopf bifurcations before coalescing with $\{0\}$ on B_{s1}. A fuller description of the bifurcations, including those occurring on B_{s2} and B_{s3}, is provided by Holmes [16]. First consider the case where μ crosses B_{s2} from region I to region III, not at 0. Here the eigenvalues indicate that a saddle-node bifurcation occurs. In Holmes [16] exact expressions are derived for the new fixed points $\{\pm x_0\}$ in the two mode case. This then approximates the behaviour of the full evolution equation and the associated semiflow $F_t^{\mu} : X \to X$ and we can thus assert that a symmetric saddle-node bifurcation occurs on a one dimensional manifold as shown in figure 1 and that the "new" fixed points are sinks in region III. Next consider μ crossing $B_h \setminus 0$. Here the eigenvalue evolution shows that a Hopf bifurcation occurs on a two-manifold and use of the stability calculations from Marsden and M^c^Cracken [32]† indicate that the family of closed orbits existing in region II are attracting.

Now let μ cross $B_{s2} \setminus 0$ from region II to region IIIa. Here the closed orbits presumably persist, since they lie at a finite distance from the bifurcating fixed point $\{0\}$. In fact the new points $\{\pm x_0\}$ appearing on B_{s2} are saddles in region IIIa, with two eigenvalues of spectrum $DG_{\mu}(\pm x_0)$ outside the unit circle and all others within it $((\lambda > 1) = 2)$. As this bifurcation occurs one of

† This has been confirmed for eight and twelve mode models by B.Hassard using his and Wan's stability formula (to appear in J. Math. An. Appl.)

the eigenvalues of spectrum $DF_t^{\mu}(0)$ passes into the unit circle so that throughout regions IIIa and III $(\lambda > 1) = 1$ for $\{0\}$. Finally consider what happens when μ crosses B_h' from region IIIa to III. Here $\{\pm x_0\}$ undergo simultaneous Hopf bifurcations and the stability calculations show that the resultant sinks in region III are surrounded by a family of repelling closed orbits. We do not yet know how the multiple closed orbits of region III interact or whether any other bifurcations occur but we now have a partial picture of behaviour near 0 derived from the two-mode approximation and from use of the stability criterion. The key to completing this analysis lies in the point 0, the "organizing centre" of the bifurcation set at which B_{s2}, B_h, and B_h' meet.

According to our general scheme, we now postulate that our bifurcation diagram near 0 is stable to small perturbations in our (approximate) equations. We look in Takens' classification and find that exactly one of them is consistent with the information found in Figure 16, namely the one shown in Figure 10. Thus we are led to the complete bifurcation diagram shown in Figure 18 with the oscillations in various regions as shown in Figure 10.

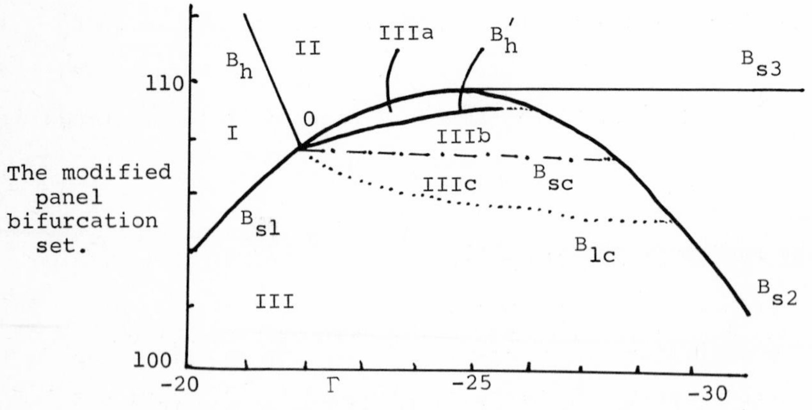

Figure 18. A local model for bifurcations of the panel near 0, $(\rho,\Gamma)\cong(110, -22.6)$; $\alpha=0.005$, $\delta=0.1$. (Numerical values derived from two mode model). For vectorfields in Regions I-IIIc, see Figure 10.

In principle one could check this out rigorously by proving that our vector field on the center manifold has the appropriate normal form. Such a calculation is probably rather long, but possible. See Holmes [16] and Marsden [18,19] for additional comments. Also, it is not clear how the presence of a small imperfection or static pressure differential would affect the symmetric vectorfields of Figure 10 .

Although the eigenvalue computations used in this analysis were derived from two and four models (in which A_μ of eqn (2) is replaced by a 4×4 or 8×8 matrix and X is replaced by a vector space isomorphic to $\mathbb{R}^4$ or $\mathbb{R}^8$) , the convergence estimates of [18,19] indicate that in the infinite dimensional case the behaviour remains qualitatively identical. In particular, for $\mu \in U$, a neighbourhood of 0 , all eigenvalues but two remain in the negative half-plane. Thus the dimension of the center manifold does not increase and our four dimensional "essential model" , a two parameter vectorfield on a two manifold, provides a local model for the onset of flutter and divergence. We are therefore justified in locally replacing the infinite dimensional semi-flow $F_t^\mu : X \to X$ by a finite dimensional system. Moreover, the actual vectorfields and bifurcation set shown in figure 10 can be realised by the nonlinear oscillator

$$\ddot{y} + \lambda_2\dot{y} + \lambda_1 y + \gamma y^2\dot{y} + \eta y^3 = 0 \qquad ; \gamma, \eta > 0$$

or

$$\dot{y}_1 = y_2$$

$$\dot{y}_2 = -\lambda_1 y_1 - \lambda_2 y_2 - \gamma y_1^2 y_2 - \eta y_1^3 \qquad (5)$$

(see Holmes and Rand [23] for a complete analysis of this system.)

In engineering terms (5) might be thought of as a "nonlinear normal mode" [44] of the system of equation (1), with λ_1, λ_2 representing equivalent linear stiffness and damping. (Note however that the relationship between the coordinates y_1, y_2 and any conveniently chosen basis in the function space X is likely to be nonlinear: in particular, a single "natural" normal mode model of the panel flutter problem cannot exhibit flutter, although it can diverge (see Holmes [16]) ; flutter occurs through coupling between the natural (linear) normal modes.)

We have thus seen how a simple nonlinear oscillator of van der Pol-Duffing type might provide an essential model for panel flutter. The methods outlined in this article may be useful in many other complex problems involving nonlinear oscillations.

REFERENCES

[1] A.A. Andronov, E.A. Leontovich, I.I Gordon and A.G. Maier "Theory of Bifurcations of Dynamic Systems on a Plane", Wiley (1971)

[2] V. Arnold, Bifurcations in Versal Families, Russian Math. Surveys, 27 (1972) 54-123

[3] R.L. Bisplinghoff and H. Ashley, "Principles of Aeroelasticity", Wiley (1962)

[4] T. Brooke Benjamin, Dynamics of a System of Articulated Pipes Conveying Fluid, I. Theory Proc. Roy. Soc. London A261(1961) 457-486 .

[5] ---------------, Dynamics of a System of Articulated Pipes Conveying Fluid, II. Experiments, Proc. Roy. Soc. London A261 (1961) 487-499.

[6] M. Buchner, J. Marsden and S. Schecter, A Differential Topology Approach to Bifurcation at Multiple Eigenvalues (to appear)

[7] L. Cesari, Functional Analysis, Nonlinear Differential Equations and the Alternative Method, in "Nonlinear Functional Analysis and Differential Equations" ed. L. Cesari, R. Kannan and J. Schuur Marcel Dekker, New York, 1976.

[8] S. Chow, J. Hale and J. Mallet-Paret, Applications of Generic Bifurcation, II, Arch. Rat. Mech. Anal. 62(1976) 209-236.

[9] M. Crandall and P. Rabinowitz, Bifurcation from Simple Eigenvalues, J. Funct. Anal. 8(1971), 321-340.

[10] __________, Bifurcation, Perturbation of Simple Eigenvalues and Linearized Stability, Arch. Rat. Mech. An. 52(1973), 161-180.

[11] E.H. Dowell, "Aeroelasticity of Plates and Shells", Noordhoff, (1975).

[12] __________, Nonlinear Oscillations of a Fluttering Plate. AIAA Vol. 4(1966), 1267-1275.

[13] A. Fischer, J. Marsden and V. Moncrief, Linearization Stability of the Einstein Equations (in preparation).

[14] Y.C. Fung, "An Introduction to the Theory of Aeroelasticity", Wiley (1955).

[15] C. Hayashi, "Nonlinear Oscillations in Physical Systems," McGraw-Hill (1964).

[16] P. Holmes, Bifurcations to Divergence and Flutter in Flow Induced Oscillations - A Finite Dimensional Analysis, J. Sound and Vib. 53(4) (1977), 471-503.

[17] P. Holmes and Y.K. Lin, Deterministic Stability Analysis of a Wind Loaded Structure, Trans. A.S.M.E. J. Appl. Mech.(to appear).

[18] P. Holmes and J. Marsden, Bifurcations to Divergence and Flutter in Flow Induced Oscillations - an Infinite Dimensional Analysis, Proc. 2nd IFAC Symposium on Distributed Parameter Systems, Warwick, 1977 (to appear).

[19] ____________, Qualitative Analysis of a Class of Partial Differential Equations Describing Flutter (in preparation).

[20] P. Holmes and D. Rand, Identification of Vibrating Systems by Generic Modelling, Proc. Intern. Symp. on Shipboard Acoustics, Delft, ed. by J.H. Janssen, Elsevier (1977).

[21] ____________, The Bifurcations of Duffing's Equation: An Application of Catastrophe Theory, J. Sound and Vib. 44(1976), 237-253.

[22] ____________, Bifurcations of the Forced van der Pol Oscillator. Quart. Appl. Math. (to appear).

[23] ____________, Bifurcations of a Nonlinear Oscillator: Free Oscillations. University of Warwick Mathematics Institute preprint (1977).

[24] W.D. Ivan and R.D. Blevins, A Model for Vortex Induced Oscillation of Structures. Trans. A.S.M.E. J. Appl. Mech. 41 (1974), 581-586.

[25] J.B. Keller and S. Antman, "Bifurcation Theory and Nonlinear Eigenvalue Problems", Benjamin (1969).

[26] J.P. Keener and H.B. Keller, Perturbed Bifurcation Theory, Arch. Rat. Mech. An. 50 (1973), 159-175.

[27] G.H. Knightly and D. Sather, Existence and Stability of Axisymmetric Buckled States of Spherical Shells, Arch. Rat. Mech. An. 63 (1977), 305-320.

[28] R. Landl, A Mathematical Model for Vortex - Excited Vibrations of Bluff Bodies, Journ. Sound and Vib. 42 (1975), 219-234.

[29] Y.K. Lin and P.J. Holmes, Stochastic Analysis of a Wind Loaded Structure , Proc. A.S.C.E. J. Engng. Mech. (to appear).

[30] J.B. McLeod and D.H. Sattinger, Loss of Stability and Bifurcation at a Double Eigenvalue, Journ. Funct. Anal. 14 (1973), 62-84.

[31] M.G. Marietta, An Isoperimetric Problem for Continuous Systems: The Aeolian Vibration of a Conductor Span. J. Franklin Inst. 301 (4) (1976), 317-333.

[32] J. Marsden and M. McCracken, "The Hopf Bifurcation and its applications" Applied Math. Sciences, #19, Springer (1976).

[33] E. Mettler, Dynamic Buckling, in "Handbook of Engineering Mechanics", ed. S. Flügge, McGraw-Hill (1962).

[34] E. Naudascher (ed.), "Flow Induced Structural Vibrations" (Proc. I.U.T.A.M.-I.A.H.R. Symposium, Karlsruhe, 1972), Springer-Verlag (1974).

[35] L. Nirenberg, "Topics in Nonlinear Analysis", Courant Institute Lecture Notes (1974).

[36] M. Novak, Aeroelastic Galloping of Prismatic Bodies, Proc. A.S A.S.C.E. J. Eng. Mech. , February (1969), 115-142.

[37] M.P. Paidoussis, Dynamics of flexible slender cylinders in axial flow, J. Fluid Mech. 26 (1966), 717-736, 737-751.

[38] M.P. Paidoussis and N.Y. Issid, Dynamic Stability of pipes conveying fluid, J. Sound and Vibration, 33 (1974), 267-294.

[39] P.C. Parks, A stability criterion for a panel flutter problem via the second method of Liapunov, in "Differential Equations and Dynamical Systems," J.K. Hale and J.P. LaSalle (eds.), Academic Press. New York (1966).

[40] R.H. Plaut and K. Huseyin, Instability of fluid-conveying pipes under axial load, Trans. ASME, J. of Appl. Mech 42 (1975), 889–890.

[41] J. Roorda, Stability of structures with small imperfections, J. Eng. Mech. 91 (1965), 87-106, 93 (1967), 37-48 , J. Mech. Physics Solids 13 (1965), 267-280.

[42] ________, An experience in equilibrium and stability, Colloq. Intern. Symp. Buenos Aires (1971).

[43] ________, On the Buckling of Symmetric Structural Systems with First and Second Order Imperfections, Int. J. Solids Structures 4 (1968), 1137-1148.

[44] R.M. Rosenberg, On Nonlinear Vibrations of Systems with Many Degrees of Freedom, Adv. in Appl. Mech. 9, 159-242, Academic Press (1966).

[45] D. Sather, Branching of Solutions of Nonlinear Equations, Rocky Mountain J. 3 (1973), 203-250.

[46] ________, Bifurcation and Stability for a Class of Shells, Arch. Rat. Mech. An. 63 (1977), 295-304.

[47] D. Sattinger, "Topics in Stability and Bifurcation Theory," Springer-Verlag Lecture Notes #309 (1973).

[48] I. Segal, Nonlinear Semigroups, Ann. of Math. 78 (1963), 339-364.

[49] A.O. St. Hilaire, Analytical Prediction of the Non-Linear Response of a Self-Excited Structure. J. Sound and Vibration 47 (2) (1976), 185-205.

[50] J. Sotomayor, Generic One Parameter Families of Vector Fields on Two Dimensional Manifolds, Publ. I.H.E.S. 43 (1974), 5-46.

[51] F. Takens, Unfoldings of Certain Singularities of Vector Fields: Generalised Hopf Bifurcations. J. Diff. Eng. 14 (1973), 476-493.

[52] ________, Singularities of Vector Fields, Publ. I.H.E.S. 43 (1974), 47-100.

[53] ________, Introduction to Global Analysis, Comm. #2, Math, Inst. Utrecht (1974).

[54] ________, Forced Oscillations and Bifurcations, Comm. #3, Math. Inst. Utrecht (1974).

[55] R. Thom, Stabilité Structurelle et Morphohénèse, Benjamin, (1972) [English translation by D.H. Fowler, 1975].

[56] J.M.T. Thompson and G.W. Hunt, "A General Theory of Elastic Stability", Wiley (1973).

[57] S. Timoshenko et al., "Vibration Problems in Engineering", 4th Ed. Wiley (1974).

[58] R. Vacaitis, M. Shinozuka and M. Takeno, Parametric Study of Wind Loading on Structures, Proc. A.S.C.E. J. Structural Division 3 (1973), 453-468.

[59] E.C. Zeeman, Euler Buckling, in "Structural Stability, the theory of Catastrophes, and Applications in the Sciences", Springer-Verlag Lecture Notes in Mathematics #525 (1976), 373-395.

[60] H. Zeigler, "Principles of Structural Stability," Ginn-Blaisdell (1968).

Vol. 489: J. Bair and R. Fourneau, Etude Géométrique des Espaces Vectoriels. Une Introduction. VII, 185 pages. 1975.

Vol. 490: The Geometry of Metric and Linear Spaces. Proceedings 1974. Edited by L. M. Kelly. X, 244 pages. 1975.

Vol. 491: K. A. Broughan, Invariants for Real-Generated Uniform Topological and Algebraic Categories. X, 197 pages. 1975.

Vol. 492: Infinitary Logic: In Memoriam Carol Karp. Edited by D. W. Kueker. VI, 206 pages. 1975.

Vol. 493: F. W. Kamber and P. Tondeur, Foliated Bundles and Characteristic Classes. XIII, 208 pages. 1975.

Vol. 494: A Cornea and G. Licea. Order and Potential Resolvent Families of Kernels. IV, 154 pages. 1975.

Vol. 495: A. Kerber, Representations of Permutation Groups II. V, 175 pages. 1975.

Vol. 496: L. H. Hodgkin and V. P. Snaith, Topics in K-Theory. Two Independent Contributions. III, 294 pages. 1975.

Vol. 497: Analyse Harmonique sur les Groupes de Lie. Proceedings 1973–75. Edité par P. Eymard et al. VI, 710 pages. 1975.

Vol. 498: Model Theory and Algebra. A Memorial Tribute to Abraham Robinson. Edited by D. H. Saracino and V. B. Weispfenning. X, 463 pages. 1975.

Vol. 499: Logic Conference, Kiel 1974. Proceedings. Edited by G. H. Müller, A. Oberschelp, and K. Potthoff. V, 651 pages 1975.

Vol. 500: Proof Theory Symposion, Kiel 1974. Proceedings. Edited by J. Diller and G. H. Müller. VIII, 383 pages. 1975.

Vol. 501: Spline Functions, Karlsruhe 1975. Proceedings. Edited by K. Böhmer, G. Meinardus, and W. Schempp. VI, 421 pages. 1976.

Vol. 502: János Galambos, Representations of Real Numbers by Infinite Series. VI, 146 pages. 1976.

Vol. 503: Applications of Methods of Functional Analysis to Problems in Mechanics. Proceedings 1975. Edited by P. Germain and B. Nayroles. XIX, 531 pages. 1976.

Vol. 504: S. Lang and H. F. Trotter, Frobenius Distributions in GL_2-Extensions. III, 274 pages. 1976.

Vol. 505: Advances in Complex Function Theory. Proceedings 1973/74. Edited by W. E. Kirwan and L. Zalcman. VIII, 203 pages. 1976.

Vol. 506: Numerical Analysis, Dundee 1975. Proceedings. Edited by G. A. Watson. X, 201 pages. 1976.

Vol. 507: M. C. Reed, Abstract Non-Linear Wave Equations. VI, 128 pages. 1976.

Vol. 508: E. Seneta, Regularly Varying Functions. V, 112 pages. 1976.

Vol. 509: D. E. Blair, Contact Manifolds in Riemannian Geometry. VI, 146 pages. 1976.

Vol. 510: V. Poènaru, Singularités C^∞ en Présence de Symétrie. V, 174 pages. 1976.

Vol. 511: Séminaire de Probabilités X. Proceedings 1974/75. Edité par P. A. Meyer. VI, 593 pages. 1976.

Vol. 512: Spaces of Analytic Functions, Kristiansand, Norway 1975. Proceedings. Edited by O. B. Bekken, B. K. Øksendal, and A. Stray. VIII, 204 pages. 1976.

Vol. 513: R. B. Warfield, Jr. Nilpotent Groups. VIII, 115 pages. 1976.

Vol. 514: Séminaire Bourbaki vol. 1974/75. Exposés 453 – 470. IV, 276 pages. 1976.

Vol. 515: Bäcklund Transformations. Nashville, Tennessee 1974. Proceedings. Edited by R. M. Miura. VIII, 295 pages. 1976.

Vol. 516: M. L. Silverstein, Boundary Theory for Symmetric Markov Processes. XVI, 314 pages. 1976.

Vol. 517: S. Glasner, Proximal Flows. VIII, 153 pages. 1976.

Vol. 518: Séminaire de Théorie du Potentiel, Proceedings Paris 1972–1974. Edité par F. Hirsch et G. Mokobodzki. VI, 275 pages. 1976.

Vol. 519: J. Schmets, Espaces de Fonctions Continues. XII, 150 pages. 1976.

Vol. 520: R. H. Farrell, Techniques of Multivariate Calculation. X, 337 pages. 1976.

Vol. 521: G. Cherlin, Model Theoretic Algebra – Selected Topics. IV, 234 pages. 1976.

Vol. 522: C. O. Bloom and N. D. Kazarinoff, Short Wave Radiation Problems in Inhomogeneous Media: Asymptotic Solutions. V. 104 pages. 1976.

Vol. 523: S. A. Albeverio and R. J. Høegh-Krohn, Mathematical Theory of Feynman Path Integrals. IV, 139 pages. 1976.

Vol. 524: Séminaire Pierre Lelong (Analyse) Année 1974/75. Edité par P. Lelong. V, 222 pages. 1976.

Vol. 525: Structural Stability, the Theory of Catastrophes, and Applications in the Sciences. Proceedings 1975. Edited by P. Hilton. VI, 408 pages. 1976.

Vol. 526: Probability in Banach Spaces. Proceedings 1975. Edited by A. Beck. VI, 290 pages. 1976.

Vol. 527: M. Denker, Ch. Grillenberger, and K. Sigmund, Ergodic Theory on Compact Spaces. IV, 360 pages. 1976.

Vol. 528: J. E. Humphreys, Ordinary and Modular Representations of Chevalley Groups. III, 127 pages. 1976.

Vol. 529: J. Grandell, Doubly Stochastic Poisson Processes. X, 234 pages. 1976.

Vol. 530: S. S. Gelbart, Weil's Representation and the Spectrum of the Metaplectic Group. VII, 140 pages. 1976.

Vol. 531: Y.-C. Wong, The Topology of Uniform Convergence on Order-Bounded Sets. VI, 163 pages. 1976.

Vol. 532: Théorie Ergodique. Proceedings 1973/1974. Edité par J.-P. Conze and M. S. Keane. VIII, 227 pages. 1976.

Vol. 533: F. R. Cohen, T. J. Lada, and J. P. May, The Homology of Iterated Loop Spaces. IX, 490 pages. 1976.

Vol. 534: C. Preston, Random Fields. V, 200 pages. 1976.

Vol. 535: Singularités d'Applications Differentiables. Plans-sur-Bex. 1975. Edité par O. Burlet et F. Ronga. V, 253 pages. 1976.

Vol. 536: W. M. Schmidt, Equations over Finite Fields. An Elementary Approach. IX, 267 pages. 1976.

Vol. 537: Set Theory and Hierarchy Theory. Bierutowice, Poland 1975. A Memorial Tribute to Andrzej Mostowski. Edited by W. Marek, M. Srebrny and A. Zarach. XIII, 345 pages. 1976.

Vol. 538: G. Fischer, Complex Analytic Geometry. VII, 201 pages. 1976.

Vol. 539: A. Badrikian, J. F. C. Kingman et J. Kuelbs, Ecole d'Eté de Probabilités de Saint Flour V-1975. Edité par P.-L. Hennequin. IX, 314 pages. 1976.

Vol. 540: Categorical Topology, Proceedings 1975. Edited by E. Binz and H. Herrlich. XV, 719 pages. 1976.

Vol. 541: Measure Theory, Oberwolfach 1975. Proceedings. Edited by A. Bellow and D. Kölzow. XIV, 430 pages. 1976.

Vol. 542: D. A. Edwards and H. M. Hastings, Čech and Steenrod Homotopy Theories with Applications to Geometric Topology. VII, 296 pages. 1976.

Vol. 543: Nonlinear Operators and the Calculus of Variations, Bruxelles 1975. Edited by J. P. Gossez, E. J. Lami Dozo, J. Mawhin, and L. Waelbroeck, VII, 237 pages. 1976.

Vol. 544: Robert P. Langlands, On the Functional Equations Satisfied by Eisenstein Series. VII, 337 pages. 1976.

Vol. 545: Noncommutative Ring Theory. Kent State 1975. Edited by J. H. Cozzens and F. L. Sandomierski. V, 212 pages. 1976.

Vol. 546: K. Mahler, Lectures on Transcendental Numbers. Edited and Completed by B. Diviš and W. J. Le Veque. XXI, 254 pages. 1976.

Vol. 547: A. Mukherjea and N. A. Tserpes, Measures on Topological Semigroups: Convolution Products and Random Walks. V, 197 pages. 1976.

Vol. 548: D. A. Hejhal, The Selberg Trace Formula for PSL $(2, \mathbb{R})$. Volume I. VI, 516 pages. 1976.

Vol. 549: Brauer Groups, Evanston 1975. Proceedings. Edited by D. Zelinsky. V, 187 pages. 1976.

Vol. 550: Proceedings of the Third Japan – USSR Symposium on Probability Theory. Edited by G. Maruyama and J. V. Prokhorov. VI, 722 pages. 1976.

Vol. 551: Algebraic K-Theory, Evanston 1976. Proceedings. Edited by M. R. Stein. XI, 409 pages. 1976.

Vol. 552: C. G. Gibson, K. Wirthmüller, A. A. du Plessis and E. J. N. Looijenga. Topological Stability of Smooth Mappings. V, 155 pages. 1976.

Vol. 553: M. Petrich, Categories of Algebraic Systems. Vector and Projective Spaces, Semigroups, Rings and Lattices. VIII, 217 pages. 1976.

Vol. 554: J. D. H. Smith, Mal'cev Varieties. VIII, 158 pages. 1976.

Vol. 555: M. Ishida, The Genus Fields of Algebraic Number Fields. VII, 116 pages. 1976.

Vol. 556: Approximation Theory. Bonn 1976. Proceedings. Edited by R. Schaback and K. Scherer. VII, 466 pages. 1976.

Vol. 557: W. Iberkleid and T. Petrie, Smooth S^1 Manifolds. III, 163 pages. 1976.

Vol. 558: B. Weisfeiler, On Construction and Identification of Graphs. XIV, 237 pages. 1976.

Vol. 559: J.-P. Caubet, Le Mouvement Brownien Relativiste. IX, 212 pages. 1976.

Vol. 560: Combinatorial Mathematics, IV, Proceedings 1975. Edited by L. R. A. Casse and W. D. Wallis. VII, 249 pages. 1976.

Vol. 561: Function Theoretic Methods for Partial Differential Equations. Darmstadt 1976. Proceedings. Edited by V. E. Meister, N. Weck and W. L. Wendland. XVIII, 520 pages. 1976.

Vol. 562: R. W. Goodman, Nilpotent Lie Groups: Structure and Applications to Analysis. X, 210 pages. 1976.

Vol. 563: Séminaire de Théorie du Potentiel. Paris, No. 2. Proceedings 1975–1976. Edited by F. Hirsch and G. Mokobodzki. VI, 292 pages. 1976.

Vol. 564: Ordinary and Partial Differential Equations, Dundee 1976. Proceedings. Edited by W. N. Everitt and B. D. Sleeman. XVIII, 551 pages. 1976.

Vol. 565: Turbulence and Navier Stokes Equations. Proceedings 1975. Edited by R. Temam. IX, 194 pages. 1976.

Vol. 566: Empirical Distributions and Processes. Oberwolfach 1976. Proceedings. Edited by P. Gaenssler and P. Révész. VII, 146 pages. 1976.

Vol. 567: Séminaire Bourbaki vol. 1975/76. Exposés 471–488. IV, 303 pages. 1977.

Vol. 568: R. E. Gaines and J. L. Mawhin, Coincidence Degree, and Nonlinear Differential Equations. V, 262 pages. 1977.

Vol. 569: Cohomologie Etale SGA $4\frac{1}{2}$. Séminaire de Géométrie Algébrique du Bois-Marie. Edité par P. Deligne. V, 312 pages. 1977.

Vol. 570: Differential Geometrical Methods in Mathematical Physics, Bonn 1975. Proceedings. Edited by K. Bleuler and A. Reetz. VIII, 576 pages. 1977.

Vol. 571: Constructive Theory of Functions of Several Variables, Oberwolfach 1976. Proceedings. Edited by W. Schempp and K. Zeller. VI. 290 pages. 1977

Vol. 572: Sparse Matrix Techniques, Copenhagen 1976. Edited by V. A. Barker. V, 184 pages. 1977.

Vol. 573: Group Theory, Canberra 1975. Proceedings. Edited by R. A. Bryce, J. Cossey and M. F. Newman. VII, 146 pages. 1977.

Vol. 574: J. Moldestad, Computations in Higher Types. IV, 203 pages. 1977.

Vol. 575: K-Theory and Operator Algebras, Athens, Georgia 1975. Edited by B. B. Morrel and I. M. Singer. VI, 191 pages. 1977.

Vol. 576: V. S. Varadarajan, Harmonic Analysis on Real Reductive Groups. VI, 521 pages. 1977.

Vol. 577: J. P. May, E_∞ Ring Spaces and E_∞ Ring Spectra. IV, 268 pages. 1977.

Vol. 578: Séminaire Pierre Lelong (Analyse) Année 1975/76. Edité par P. Lelong. VI, 327 pages. 1977.

Vol. 579: Combinatoire et Représentation du Groupe Symétrique, Strasbourg 1976. Proceedings 1976. Edité par D. Foata. IV, 339 pages. 1977.

Vol. 580: C. Castaing and M. Valadier, Convex Analysis and Measurable Multifunctions. VIII, 278 pages. 1977.

Vol. 581: Séminaire de Probabilités XI, Université de Strasbourg. Proceedings 1975/1976. Edité par C. Dellacherie, P. A. Meyer et M. Weil. VI, 574 pages. 1977.

Vol. 582: J. M. G. Fell, Induced Representations and Banach *-Algebraic Bundles. IV, 349 pages. 1977.

Vol. 583: W. Hirsch, C. C. Pugh and M. Shub, Invariant Manifolds. IV, 149 pages. 1977.

Vol. 584: C. Brezinski, Accélération de la Convergence en Analyse Numérique. IV, 313 pages. 1977.

Vol. 585: T. A. Springer, Invariant Theory. VI, 112 pages. 1977.

Vol. 586: Séminaire d'Algèbre Paul Dubreil, Paris 1975–1976 (29ème Année). Edited by M. P. Malliavin. VI, 188 pages. 1977.

Vol. 587: Non-Commutative Harmonic Analysis. Proceedings 1976. Edited by J. Carmona and M. Vergne. IV, 240 pages. 1977.

Vol. 588: P. Molino, Théorie des G-Structures: Le Problème d'Equivalence. VI, 163 pages. 1977.

Vol. 589: Cohomologie l-adique et Fonctions L. Séminaire de Géométrie Algébrique du Bois-Marie 1965–66, SGA 5. Edité par L. Illusie. XII, 484 pages. 1977.

Vol. 590: H. Matsumoto, Analyse Harmonique dans les Systèmes de Tits Bornologiques de Type Affine. IV, 219 pages. 1977.

Vol. 591: G. A. Anderson, Surgery with Coefficients. VIII, 157 pages. 1977.

Vol. 592: D. Voigt, Induzierte Darstellungen in der Theorie der endlichen, algebraischen Gruppen. V, 413 Seiten. 1977.

Vol. 593: K. Barbey and H. König, Abstract Analytic Function Theory and Hardy Algebras. VIII, 260 pages. 1977.

Vol. 594: Singular Perturbations and Boundary Layer Theory, Lyon 1976. Edited by C. M. Brauner, B. Gay, and J. Mathieu. VIII, 539 pages. 1977.

Vol. 595: W. Hazod, Stetige Faltungshalbgruppen von Wahrscheinlichkeitsmaßen und erzeugende Distributionen. XIII, 157 Seiten. 1977.

Vol. 596: K. Deimling, Ordinary Differential Equations in Banach Spaces. VI, 137 pages. 1977.

Vol. 597: Geometry and Topology, Rio de Janeiro, July 1976. Proceedings. Edited by J. Palis and M. do Carmo. VI, 866 pages. 1977.

Vol. 598: J. Hoffmann-Jørgensen, T. M. Liggett et J. Neveu, Ecole d'Eté de Probabilités de Saint-Flour VI – 1976. Edité par P.-L. Hennequin. XII, 447 pages. 1977.

Vol. 599: Complex Analysis, Kentucky 1976. Proceedings. Edited by J. D. Buckholtz and T. J. Suffridge. X, 159 pages. 1977.

Vol. 600: W. Stoll, Value Distribution on Parabolic Spaces. VIII, 216 pages. 1977.

Vol. 601: Modular Functions of one Variable V, Bonn 1976. Proceedings. Edited by J.-P. Serre and D. B. Zagier. VI, 294 pages. 1977.

Vol. 602: J. P. Brezin, Harmonic Analysis on Compact Solvmanifolds. VIII, 179 pages. 1977.

Vol. 603: B. Moishezon, Complex Surfaces and Connected Sums of Complex Projective Planes. IV, 234 pages. 1977.

Vol. 604: Banach Spaces of Analytic Functions, Kent, Ohio 1976. Proceedings. Edited by J. Baker, C. Cleaver and Joseph Diestel. VI, 141 pages. 1977.

Vol. 605: Sario et al., Classification Theory of Riemannian Manifolds. XX, 498 pages. 1977.

Vol. 606: Mathematical Aspects of Finite Element Methods. Proceedings 1975. Edited by I. Galligani and E. Magenes. VI, 362 pages. 1977.

Vol. 607: M. Métivier, Reelle und Vektorwertige Quasimartingale und die Theorie der Stochastischen Integration. X, 310 Seiten. 1977.

Vol. 608: Bigard et al., Groupes et Anneaux Réticulés. XIV, 334 pages. 1977.